Machine Learners

Machine Learners

Archaeology of a Data Practice

Adrian Mackenzie

The MIT Press
Cambridge, Massachusetts
London, England

This book was set in Stone Serif by Westchester Publishing Services and was printed and bound in the United States of America.

Library of Congress Cataloging-in-Publication Data

Names: Mackenzie, Adrian, 1962- author.
Title: Machine learners : archaeology of a data practice / Adrian Mackenzie.
Description: Cambridge, MA : The MIT Press, [2017] | Includes bibliographical references and index.
Identifiers: LCCN 2017005343 | ISBN 9780262036825 (hardcover : alk. paper)
Subjects: LCSH: Information theory. | Machine learning–Philosophy. | Electronic data processing–Philosophy.
Classification: LCC Q360 .M3134 2017 | DDC 003/.54–dc23
LC record available at https://lccn.loc.gov/2017005343

10 9 8 7 6 5 4 3 2 1

Contents

List of Figures

List of Tables

Preface

Although this book is not an ethnography, it has an ethnographic situation. If it has a field site, then it lies close to the places where the writing was done—in universities, on campuses, in classrooms and online training courses (including massive open online causes [MOOCs]), and then amid the books, documents, websites, software manuals and documentation, and a rather vast accumulation of scientific publications. It's a case of "dig where you stand" or "auto-archaeology."

Readers familiar with textbooks in computer science and statistics can detect the traces of those fields in various typographic conventions drawn from the fields about which I write. Important conventions include:

1. Typesetting the name of any code or devices that do machine learning and datasets on which machine learners operate in a `monospace` or terminal font: `machine learner` or `iris`;
2. Presenting formulae, functions, and equations using the bristling indexicality of mathematical typography: $\hat{\beta}$

I emulate the apparatus of science and engineering publication as an experiment in *in-situ* hybridization. Social science and humanities researchers, even when they are observant participants in their field sites, rarely experience a coincidence between their own writing practices and that of the participants in the research site they study. The object of study in this book, however, is a knowledge practice that documents itself in code, equations, diagrams, and statements circulated in articles, books, and various online formats (blogs, wikis, software repositories). It is also possible for a social researcher to adopt some of these conventions, even if they sometimes take us to the limits of coherence, and of easily understandable expression.

I've been writing code for years (Mackenzie 2006). Writing code was nearly always something distant from writing about code because coding was about software projects and writing was about thinking and knowledge. I was slow to realize they are much entangled. Recent developments in ways of analyzing and publishing scientific data bring coding and writing closer together. Implementing code can be done almost in the same space, in the same screen or pane, as writing about code. The mingling of coding and writing about code brings about sometimes generative, sometimes frustrating encounters with various scientific knowledge (mathematics, statistics, computer science), with infrastructures and devices on many scales (ranging across networks, text editors, databases here and there, hardware and platforms of various kinds, as well as interfaces) and many domains.

At many points in researching the book, I digressed a long way into remote technical domains of statistical inference, probability theory, linear algebra, dynamic models, as well as database design and data standards. In the interests of connecting the many propositions, formulations, diagrams, equations, citations, and images in this book, much of the code I've written in implementing machine learning models or reconstructing certain data practices does not appear in this text, just as not all of the words I've written in trying to construct arguments or think about data practices have been included. Much has been cut away and left on the ground (although the `git` repository of the book preserves many traces of the writing and code; see https://github.com/datapractice/machinelearning). As in the many machine learning textbooks, recipe books, cookbooks, how-tos, tutorials, and manuals I have read, code, graphics, and prose have been tidied here. Many exploratory forays are lost and almost forgotten.

The several years I have spent doing and writing about data practice have felt substantially different than any other project by virtue of the hybridization between code in text and text in code. Practically, this is made possible by working on code and text within the same file, in the same text editor. Switching between writing `R` and Python code (about which I say more below) to retrieve data, to transform it, to produce graphics, to construct models or some kind of graphic image, and within the same file be writing academic prose, might be one way to write about machine learning as a data practice.

The capacity to mingle text, code, and images depends on an ensemble of open source, often command-line software tools that differ somewhat from the typical social scientist or humanities researchers' software toolkit of word processor, bibliographic software, image editor, and web browser. In particular, I have relied on software packages in the `R` programming language such as the "`knitr`" (Xie 2013; Xie and Allaire 2012) and in `python`, the `ipython` notebook environment (Perez and Granger 2007). Both have been developed by scientists and statisticians in the name of "reproducible research." Many examples of this form of writing can be found on the web: see IPython Notebook Viewer for a sample of these. These packages are designed to allow a combination of code written in `R`, python, or other programming languages, scientific writing (including mathematical formula) and images to be included and, importantly, executed together to produce a document.[1]

In making use of the equipment created by the people I study, I've attempted to bring the writing of code and writing about code-like operations into critical proximity. Does proximity or mixing of writing code and writing words make a practical

1 In order to do this, they typically combine some form of text formatting or "markup," that ranges from very simple formatting conventions (e.g., the "Markdown" format used in this book is much less complicated than HTML and uses markup conventions readable as plain text and modeled on email [Gruber 2004]) to the highly technical (LaTeX, the de-facto scientific publishing format or "document preparation system" (Lamport and LaTEX 1986) elements of which are also used here to convey mathematical expressions). They add to that blocks of code and inline code fragments that are executed as the text is formatted to produce results that are shown in the text or inserted as figures in the text.

There are a few different ways of weaving together text, computation and images together. Each suffers from different limitations. In `ipython`, a scientific computing platform dating from 2005 (Perez and Granger 2007) and used across a range of scientific settings, interactive visualization and plotting, as well as access to operating system functions are brought together in a `Python` programming environment. Especially in using the `ipython` notebook, where editing text and editing code is all done in the same window, and the results of changes to code can be seen immediately, practices of working with data can be directly woven together with writing about practice. By contrast, `knitr` generates documents by combining text passages and the results (graphs, calculations, tabulations of data) of code interleaved between the text into one output document. When `knitr` runs, it executes the code and inserts the results (calculations, text, images) in the flow of text.

Practically, this means that the text editor used to write code and text, remains somewhat separate from the software that executes the code. By contrast, `ipython` combines text and `Python` code more continuously, but at the cost of editing and writing code and text in a browser window. Most of the conveniences and affordances of text editing software is lost. While `ipython` focuses on interactive computation, `knitr` focuses on bringing together scientific document formatting and computation. Given that both can include code written in other languages (that is, `python` code can be processed by `knitr`, and `R` code executed in `ipython`), the differences are not crucially important. This whole book could have been written using just Python, since Python is a popular general purpose programming language, and many statistical, machine learning and data analysis libraries have been written for Python. I have used both, sometimes to highlight tensions between the somewhat more research-oriented `R` and the more practical applications typical of Python, and sometimes because code in one language is more easily understood than the other.

difference to an account of practice? If recent theories of code and software as forms of speech, expression or performative utterance (Cox 2012; Coleman 2012), or more generally praxiography as a reality-making descriptive practice (Mol 2003) are right, then it should. Weaving code through writing in one domain of contemporary technical practice, machine learning, might be one way of keeping multiple practices present, developing a concrete sense of abstraction, and allowing an affective expansion in relation to machines.

Acknowledgments

From 2007 to 2012, I benefited greatly from a research position in the UK Economic and Social Research Council-funded Centre for Economic and Social Aspects of Genomics at Lancaster University. Certain colleagues there, initially in the "Sociomics Core Facility," participated in the inception of this book. Ruth McNally with her almost geeky interest in genomics, Paul Oldham with his enthusiasm for "all the data," Maureen McNeil with her critical acuity, and Brian Wynne with his connective thought were participants in many discussions concerning the transformation of life sciences around which my interest in machine learning first crystallized.

The Technology in Practice Group at ITU Copenhagen hosted some of the research work during 2014. Brit Ross Winthereik in particular made it possible for us to develop some of the key ideas. I was also lucky to be part of an excellent research team on "Socialising Big Data" (2013–2015) that included Penny Harvey, Celia Lury, Ruth McNally, and Evelyn Ruppert. We had excellent discussions.

My colleagues in the science studies at Lancaster, especially Maggie Mort, Richard Tutton, Lucy Suchman, and Claire Waterton, are always a delight to work with. I have also been fortunate to have worked with inspiring and adventurous doctoral students at Lancaster during the writing of this book. Lara Houston, Mette Kragh Furbo, Felipe Raglianti, Emils Kilis, Xaroula Charalampia, and Nina Ellis have all helped and provided inspiration in different ways. Sjoerd Bollebakker kindly updated many of the scientific literature searches toward the end of the book's writing.

Various academic staff in the Department of Applied Mathematics and Statistics at Lancaster University shepherded me through postgraduate statistics training courses: Brian Francis for his course of "Data Mining," David Lucy for his course on "Bayesian Statistics," and Thomas Jakl for his course "Genomic Data Analysis." Since 2015, I've also come to know some machine learners much better through the Data Science Institute, Lancaster University.

My inestimable friends in the *Computational Cultures* editorial group have listened to and irrationally encouraged various threads of work runnning the book. I warmly acknowledge the support of Matthew Fuller, Andrew Goffey, Graham Harwood, and Olga Goriunova. Nina Wakeford encouraged me to code naively.

Far away in Australia, Anna Munster has kept our conversation around data visible. As always, Celia Roberts helped me sort out what I really want to do.

1 Introduction: Into the Data

Definition: A computer program is said to **learn** from experience E with respect to some class of tasks T and performance measure P, if its performance at tasks in T, improves with experience E (Mitchell 1997, 2).

In the past fifteen years, the growth in algorithmic modeling applications and methodology has been rapid. It has occurred largely outside statistics in a new community—often called machine learning—that is mostly young computer scientists. The advances, particularly over the last five years, have been startling (Breiman 2001b, 200).

The key question isn't "How much will be automated?" It's how we'll conceive of whatever *can't* be automated at a given time (Lanier 2013, 77).

A relatively new field of scientific-engineering devices said to "learn from experience" has become operational in the last three decades. Known by various names—machine learning, pattern recognition, knowledge discovery, data mining—the field and its devices, which all take shape as computer programs or code, seem to have quickly spread across scientific disciplines, business and commercial settings, industry, engineering, media, entertainment, and government. Heavily dependent on computation, machine learners are found in breast cancer research, autonomous vehicles, insurance risk modeling, credit transaction processing, computer gaming, face and handwriting recognition systems, astronomy, advanced prosthetics, ornithology, finance, surveillance (see the U.S. government's `SkyNet` for one example of a machine learning surveillance system (National Security Agency 2012)), or robots (see a Google robotic arm farm learning to sort drawers of office equipment such as staplers, pens, erasers and paper clips (Levine et al. 2016)).

Sometimes machine learning devices are understood as *scientific models*, and sometimes they are understood as *operational algorithms*. In many scientific fields, publications mention or describe these techniques as part of their analysis of some experimental or observational data (as in the logistic regression classification models found in many biomedical papers). They anchor the field of "data science" (Schutt and O'Neil 2013) as institutionalized in several hundred data science institutes scattered worldwide. Not so recently, they also became mundane mechanisms deeply embedded in other systems or gadgets (as in the decision tree models used in some computer game consoles to recognize gestures, the neural networks used to recognize voice commands by search engine services such as `Google Search` and `Apple Siri` (McMillan 2013), or Google's `TensorFlow` software packages that put deep convolutional neural nets on Android devices (Google 2015)). In platform settings, they operate behind the scenes as part of the everyday functioning of services ranging from player ranking in online games to border control face recognition, from credit scores to news feeds on Facebook.

In all of these settings, applications, and fields, machine learning is said to transform the nature of knowledge. Might it transform the practice of critical thought? This book is an experiment in such practice.

Three accumulations: Settings, data, and devices

Three different accumulations cross-stratify in machine learning: settings, data, and devices. The volume and geography of searches on Google Search provide some evidence of the diverse settings or sites doing machine learning. If we search for terms such as `artificial intelligence`, `machine learning`, and `data mining` on the Google Trends service, the results for the last decade or so suggest shifting interest in these topics.

In figure 1.1, two general search terms that had a high search volume in 2004—"artificial intelligence" and "data mining"—slowly decline over the years before starting to increase again in the last few years. By contrast, `machine learning` loses volume until around 2008 and then gradually rises again so that by mid-2016 it exceeds the long-standing interests in data mining and artificial intelligence. Whatever the uncertainties in understanding GoogleTrends results, these curves suggest an accumulation of sites and settings turning toward machine learning.[1] What does it mean that machine learning surfaces in so many different places, from fMRIs to Facebook's

1 In the plot (figure 1.1), the weekly variations in search volume on Google give rise to many spikes in the data. These spikes can be linked to specific events such as significant press releases, public debates, media attention, and film releases. It is hard to know who is doing these searches. The data provided by Google Trends include geography, and it would be interesting to compare the geographies of interest in the different terms over time.

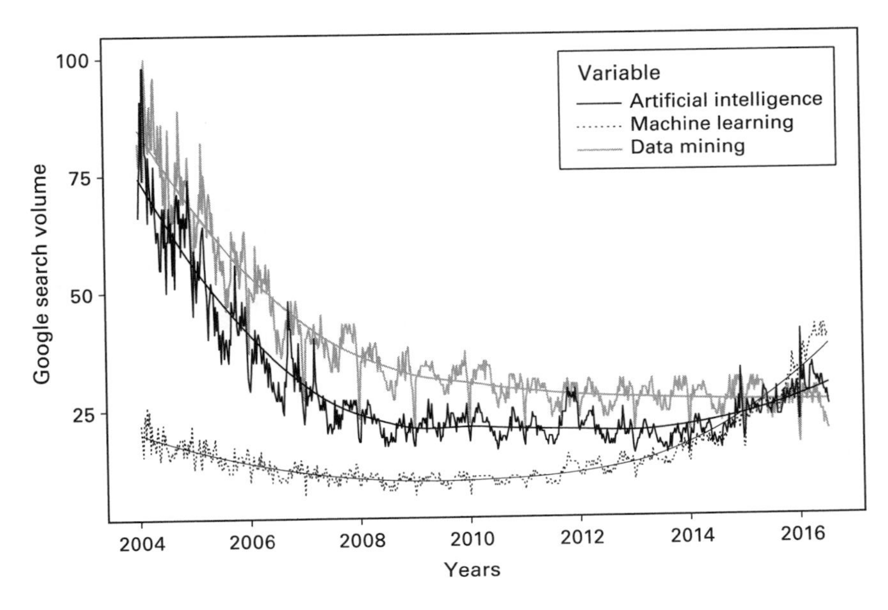

Figure 1.1
Google Trends search volume for "machine learning" and related query terms in English, globally 2004–2016.

`AI-Flow` (Facebook 2016), from fisheries management to Al Queda courier network monitoring by `SkyNet`?[2]

A second accumulation concerns the plenitude of things in the world as data. If we wanted to describe the general horizon of machine learning as a data practice in its specificity, then we might turn to cats. Cat images accumulate on websites and social media platforms as decentered, highly repetitive data forms. Like the billions of

2 The diagram shown in figure 1.1 actually draws two lines for each trend. The "raw" weekly GoogleTrends data—definitely not raw data, as it has been normalized to a percentage (Gitelman 2013)—appear in the spiky lines, but a much smoother line shows the general trend. This smoothing line is the work of a statistical model—a local regression or loess model (Cleveland, Grosse, and Shyu 1992) developed in the late 1970s. The line depends on intensive computation and models (linear regression, *k* nearest neighbors). The smoother lines make the spiky weekly search counts supplied by Google much easier to see. They construct alignments in the data by replacing the irregular variations with a curve that unequivocally runs through time with greater regularity. The smoothed lines shade the diagram with a predictive pattern. The lineaments of machine learning already appear in such lines. How have things been arranged so that smooth lines run through an accumulated archive of search data?

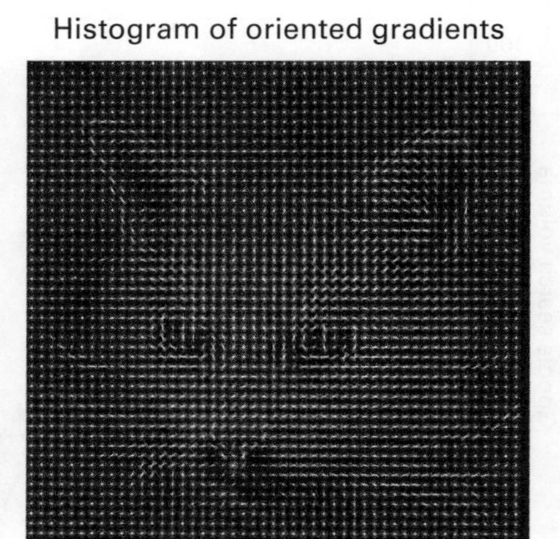

Figure 1.2
Close up of cat. The image on the left is already signal-processed as a JPEG format file. The image on the right is further processed using histogram of oriented gradients (HOG) edge detection. `kittydar` models HOG features. Cat photo courtesy photos-public-domain.com.

search engine queries, email messages, tweets, or much contemporary scientific data (e.g., the DNA sequence data discussed in chapter 7), images accumulate in archives of communication. Take the case of `kittydar`, a machine learner in the area of image recognition (see kittydar): "Kittydar is short for 'kitty radar.' Kittydar takes an image (canvas) and locates cats in the image" (Arthur 2012). This playful piece of code demonstrates how machine learning reconfigures mundane accumulation. Heather Arthur, who developed `kittydar`, writes:

> Kittydar first chops the image up into many "windows" to test for the presence of a cat head. For each window, kittydar first extracts more tractable data from the image's data. Namely, it computes the Histogram of Orient Gradients descriptor of the image . . . This data describes the directions of the edges in the image (where the image changes from light to dark and vice versa) and what strength they are. This data is a vector of numbers that is then fed into a neural network . . . which gives a number from 0 to 1 on how likely the histogram data represents a cat. The neural network . . . has been pre-trained with thousands of photos of cat heads and their histograms, as well as thousands of non-cats. See the repo for the node training scripts (Arthur 2012).

This toy device finds cat heads in digital photographs but also exemplifies many key traits of machine learning. Large accumulations of things become *vectors* in a dataset. A dataset is used to train a typical machine learning device, a neural net, and the neural net *classifies* subsequent images probabilistically. The code for all this is "in the repo." Based on how `kittydar` locates cats, we can begin to imagine similar pattern recognition techniques in use in self-driving cars (Thrun et al. 2006), border control facial recognition systems, military robots, or wherever something seen implies something to do.

Faced with the immense accumulation of cat images on the internet, `kittydar` can do little. It only detects the presence of cats that face forward. It sometimes classifies people as cats. As Arthur's description suggests, the software finds cats by cutting the images into smaller windows. For each window, it measures a set of gradients—a spatial order of great significance in machine learning—running from light and dark and then compares these measurements to the gradients of known cat images (the so-called "training data"). The work of classification according to the simple categories of "cat" or "not cat" is given either to a neural network (as discussed in chapter 8, a typical machine learning technique and one that has recently been heavily developed by researchers at Google (Le et al. 2011), themselves working on images of cats among other things taken from Youtube videos (BBC 2012), or to a support vector machine (a technique first developed in the 1990s by researchers working at IBM; see chapter 6).

A final accumulation comprises machine learning techniques and devices. Machine learners range from the mundane to the esoteric, from code miniatures such as `kittydar` to the infrastructural sublime of computational clouds and clusters twirling through internet data streams. Like the images that `kittydar` classifies, the names of machine learning techniques and devices proliferate and accumulate in textbooks, instructional courses, website tutorials, software libraries, and code listings: linear regression, logistic regression, neural networks, linear discriminant analysis, support vector machines, k-means clustering, decision trees, *k* nearest neighbors, random forests, principal component analysis, or naive Bayes classifier, to name just some of the most commonly used. Sometimes they have proper names: `RF-ACE`, `Le-Net5`, or `C4.5`. These names refer to predictive models and computational algorithms of various ilk and provenance. Intricate data practices—normalization, regularization, cross-validation, feature engineering, feature selection, optimization—embroider datasets into shapes they can recognize. The techniques, algorithms, and models are not necessarily startling new or novel. They take shape against a background of more than a century of work in mathematics, statistics, and computer science as well as disparate scientific fields ranging from anthropology to zoology. Mathematical constructs drawn from linear algebra, differential calculus, numerical optimization, and probability theory pervade practice in the field. Machine learning is an accumulation rather than a radical transformation.

Who or what is a machine learner?

I am focusing on machine learners—a term that refers to both humans and machines or human–machine relations throughout this book—situated amid these three accumulations of settings, data, and devices. Although it is not always possible to disentangle machine learners from the databases, infrastructures, platforms, or interfaces they work through, I will argue that data practices associated with machine learning delimit a *positivity* of knowing. The term "positivity" comes from Michel Foucault's *The Archaeology of Knowledge* (Foucault 1972) and refers to specific forms of accumulation of statements grouped in a discursive practice and an operational formation. Analyzed archeologically, a positivity can be investigated and inhabited to some degree by critical thought.

Foucault attributes a lift-off effect to positivity:

> The moment at which a discursive practice achieves individuality and autonomy, the moment therefore at which a single system for the formation of statements is put into operation, or the moment at which this system is transformed, might be called the threshold of positivity (Foucault 1972, 186).

Machine learners today circulate into domains that lie far afield of the eugenic and psychology laboratories, industrial research institutes, or specialized engineering settings in which they first took shape (in some cases, such as the linear regression model or principal component analysis, more than a century ago; in others such as support vector machines or random forests, in the last two decades). If they are not exactly new and have diverse genealogies, the question is: what happen as machine learners shift from localized mathematical or engineering techniques to an everyday device that can be generalized to locate cats in digital images, the Higgs boson in particle physics experiments, or fraudulent credit card transactions? Does the somewhat unruly generalization of machine learning across different epistemic, economic, institutional settings—the pronounced uptick shown in figure 1.1—attest to a redefinition of knowledge, decision, and control, a new operational formation in which "a system is transformed"?

Algorithmic control to the machine learners?

Written in code, machine learners operate as programs or computational processes to produce statements that may take the form of numbers, graphs, and propositions (see e.g., the propositions produced by a recurrent neural net on the text of this book in the concluding chapter 9). Machine learning can also be viewed as a change in how programs, or the code that controls computer operations, are developed and operate (see chapter 2 for a more detailed discussion of this). The term "learning" in machine learning points to this change, and many machine learners emphasize it. Pedro Domingos, for instance, a computer scientist at the University of Washington, writes:

> Learning algorithms—also known as learners—are algorithms that make other algorithms. With machine learning, computers write their own programs, so we don't have to (Domingos 2015, 6).

Viewed from the perspective of control, and how control is practiced, machine learners perpetuate and epitomize the "control revolution" (Beniger 1986) that arguably has, since the late 19th century, reconfigured production, distribution, consumption, and bureaucracy by tabulating, calculating, and increasingly communicating events and operations. With the growth of digital communication networks in the form of the internet, the late 20th century entered a new crisis of control, no longer centered on logistic acceleration but on communication and knowledge. Almost all accounts of the operational power of machine learning emphasize its power to regain control of processes of communication—border flows, credit fraud, spam email, financial market prices, cancer diagnosis, targeted online adverts—processes whose unruly or transient multiplicity otherwise evades or overwhelms. On this view, `kittydar` can isolate cats amid the excessive accumulation of images on the internet because neural net learning algorithms (back-propagation, gradient descent) have written a program—"a pre-trained" neural net—during its training phase.

If a newly programmatic field of control takes shape around machine learning, how would we distinguish it from computation more generally? Recent critical research on algorithms offers one lead. In a study of border control systems, which often use machine learners to do profiling and facial recognition, Louise Amoore advocates attention to calculation and algorithms:

> Surely this must be a primary task for critical enquiry—to uncover and probe the moments that come together in the making of a calculation that will automate all future decisions. To be clear, I am not proposing some form of humanist project of proper ethical judgement, but rather calling for attention to be paid to the specific temporalities and norms of algorithmic techniques that rule out, render invisible, other potential futures (Amoore 2011).

As Amoore writes, some potential futures are being "ruled out" as calculations automate decisions. Anna Munster puts the challenge more bluntly: "prediction takes down potential" (Munster 2013). I find much to agree with here. Machine learning is a convoluted but nevertheless concrete and historically specific form of calculation (as we will see in exploring algebraic operations in chapter 3, in finding and optimizing certain mathematical functions in chapter 4 or in characterizing and shaping probability distributions in chapter 5). It works to mediate future-oriented decisions (although all too often near-future decisions such as ad-click prediction).

I am less certain about treating machine learning as automation. Learning from data, as we will see, often sidesteps and substitutes for existing ways of acting, and practices of control, and it thereby reconfigures human–machine differences. Yet the notion of automation does not capture well how this comes about. The programs that machine learners "write" are formulated as probabilistic models, as learned rules or association, and they generate predictive and classificatory statements ("this is a cat"). They render calculable some things that hitherto appeared intractable to calculation (e.g., the argument of a legal case). Predictive and classificatory calculation, with all the investment it attracts (in the form of professional lives, in the form of infrastructures, in reorganization of institutions, corporations, and governments, etc.) does rule out some and reinforce other futures.[3] If this transformed calculability is automation, then we need to understand the specific contemporary reality of automation as it takes shape in machine learning. We cannot conduct critical enquiry into how calculation will automate future decisions without putting the notions of calculation and automation into question.

Does the concept of algorithm help us identify the moments that come together in machine learning without resorting to a historical concepts of automation or calculation? In various scholarly and political debates around changes in business, media, education, health, government or science, quasi-omnipotent agency has been imputed to algorithms (Barocas, Hood, and Ziewitz 2013; Beer and Burrows 2013; Cheney-Lippold 2011; Fuller and Goffey 2012; Galloway 2004; Gillespie 2014; Neyland 2015; Pasquinelli 2014; Smith 2013; Totaro and Ninno 2014; Wilf 2013) or sometimes just "the algorithm." This growing body of work understands the power of algorithms in the social science and humanities literature in different ways, sometimes in terms of rules, sometimes as functions or mathematical abstractions, and increasingly as a located practice. General agreement exists that algorithms are powerful or at least can bear down heavily on people's lives and conduct, reconfiguring, for instance, culture as algorithmic (Hallinan and Striphas 2014).

3 As for consequences, we need only consider some of the many forms of work that have already been affected by or soon could be affected by machine learning. Postal service clerks no longer sort the mail because neural net-based handwriting recognition reads addresses on envelopes. Locomotives, cars, and trucks are already driven by machine learners, and soon driving may not be the same occupational cultural it was. Hundreds of occupational categories have to some degree or other machine learners in their near future. Carl Benedikt Frey and Michael Osborne model the chances of occupational change for 700 occupations using, aptly enough, the machine learning technique of Gaussian Processes (Frey and Osborne 2013).

Some of the critical literature on algorithms identifies abstractions as the source of their power. For instance, in his discussion of the "metadata society," Paolo Pasquinelli proposes that

> a progressive political agenda for the present is about moving at the same level of abstraction as the algorithm in order to make the patterns of new social compositions and subjectivities emerge. We have to produce new revolutionary institutions out of data and algorithms. If the abnormal returns into politics as a mathematical object, it will have to find its strategy of resistance and organisation, in the upcoming century, in a mathematical way (Pasquinelli 2015).

"Moving at the same level of abstraction as the algorithm" offers some purchase as a formulation for critical practice and for experiments in such practice. Because in mathematics, let alone critical thought, abstraction can be understood in many different ways, any direct identification of algorithms with abstraction will, however, be subject to considerations of practice. Which algorithm, what kind of abstraction, and which "mathematical way" should we focus on? Like automation and calculation, abstraction and mathematics are historically mutable. We cannot "move at the same level" without taking that mutability into account. Furthermore, given the accumulations of settings, data, and devices, there might not be any single level of abstraction to move at, only a torque and flux of different moments of abstraction at work in generalizing, classifying, circulating, and stratifying in the midst of transient and plural multiplicities.

The archaeology of operations

Given that mathematics and algorithms loom large in machine learning, how do we address their workings without preemptively ascribing potency to mathematics or algorithms? In the chapters that follow, I explore specific learning algorithms (gradient descent in chapter 4 or recursive partitioning in chapter 6) and mathematical techniques (the sigmoid function in chapter 4 or inner products in chapter 3) in some empirical and conceptual depth. Following much scholarship in science and technology studies, I maintain that attention to specificity of practices is an elementary prerequisite to understanding human–machine relations and their transformations. The archaeology of operations that I will develop combines an interest in machine learning as a form of knowledge production and a strategy of power. Like Foucault, I see no exteriority between techniques of knowledge and strategies of power ("between techniques of knowledge and strategies of power, there is no exteriority, even if they have specific roles and are linked together on the basis of their difference" (Foucault 1998, 98)).

If we understand machine learning as a data practice that reconfigures local centers of power and knowledge by redrawing human–machine relations, then differences associated with machine learners in the production of knowledge should be a focus of attention. Differences are an operative concern here because many machine learners classify things. Machine learners are often simply called "classifiers." Much of the practice of difference works in terms of categories. `Kittydar` classifies images as `cat` with some probability, but categorization and classification in machine learning occurs much more widely.[4] We might understood the importance of categories sociologically. For instance, in his account of media power, Nick Couldry highlights the importance of categories and categorization:

> *Category* is a key mechanism whereby certain types of ordered (often "ritualized") practice produce power by enacting and embodying categories that serve to mark and divide up the world in certain ways. Without *some* ordering feature of practice, such as "categories," it is difficult to connect the multiplicity of practice to the workings of power, whether in the media or in any other sphere. By understanding the work of categories, we get a crucial insight into why the social world, in spite of its massive complexity still appears to us as a *common* world (Couldry 2012, 62).

Orderings of categorical differences undergo a great deal of intensification via machine learning. Categories are often simply an existing set of classifications assumed or derived from institutionalized or accepted knowledges (e.g., the categories of customers according to gender or age). Machine learners also generate new categorical workings or mechanisms of differentiation. As we will see (e.g., in chapter 7 in relation to scientific data from genomes), machine learners invent or find new sets of categories for a particular purpose (such as cancer diagnosis or prognosis). These differentiations may or may not bring social good. The person who finds him or herself paying a higher price for an air ticket by virtue of some unknown combination of factors, including age, credit score, home address, previous travel, or educational qualifications, experiences something of the troubling effects of classificatory power.

Asymmetries in common knowledge

What can critical thought, the kind of enquiry that seeks to identify the conditions that concretely constitute what anyone can say or think or do, learn from machine learning? If we see a massive regularization of order occurring in machine learning,

4 John Cheney-Lippold offers a quite general overview of categorization work. He writes: "algorithm ultimately exercises control over us by harnessing these forces through the creation of relationships between real-world surveillance data and machines capable of making statistically relevant inferences about what that data can mean" (Cheney-Lippold 2011, 178). Much of my discussion here seeks to explore the space of "statistical inference of what that data can mean" as an operational field of knowledge production.

what is at stake in trying to think through those practices? They display moments of formalization (especially mathematical and statistical), circulation (pedagogically and operationally), generalization (propagating and proliferating in many domains and settings), and stratification (the socially, epistemically, economically, and sometimes politically or ontologically loaded reiterative enactment of categories). Understanding how a support vector machine or a random forest orders differences could change how we relate to what we see, feel, sense, hear, or think in the face of a contemporary platform such as Amazon's, which uses Association Rule Mining, an app, a passport control system that matches faces of arriving passengers with images in a database, a computer game, or a genetic test (all settings in which machine learning is likely to be operating).

Machine learners sometimes complain of the monolithic and homogenous success of machine learning. Some expert practitioners bemoan the uniformity of its applications. Jeff Hammerbacher, previously chief research scientist at Facebook, cofounder of a successful data analytics company called Cloudera, and currently working on cancer research at Mount Sinai hospital, complained about the spread of machine learning in 2011: "the best of my generation are thinking about how to make people click ads" (Vance 2011). Leaving aside debates about the ranking of "best minds" (a highly competitive and exhaustively tested set of subject positions; see chapter 8), Hammerbacher was lamenting the flourishing use of predictive analytics techniques in online platforms such as Twitter, Google, and Facebook, and on websites more generally, whether they be websites that sell things or advertising space. The mathematical skills of many PhDs from MIT, Stanford, or Cambridge were wrangling data in the interests of microtargeted advertising. As Hammerbacher observes, they were "thinking about how to make people click ads," and this "thinking" mainly took and does take the form of building predictive models that tailored the ads shown on websites to clusters of individual preferences and desires.

Hammerbacher's unhappiness with ad-click prediction unexpectedly resonates with critical responses to the use of machine learning in the digital humanities. Some versions of the digital humanities make extensive use of machine learning. To cite one example, in *Macroanalysis: Digital Methods and Literary History*, Matthew Jockers describes how he relates to one currently popular machine learning or statistical modeling technique, the topic model (itself the topic of discussion in chapter 5; see also Mohr and Bogdanov 2013):

> If the statistics are rather too complex to summarize here, I think it is fair to skip the mathematics and focus on the end results. We needn't know how long and hard Joyce sweated over *Ulysses* to appreciate his genius, and a clear understanding of the LDA machine is not required in order to see the beauty of the result (Jockers 2013, 124).

The widely used Latent Dirichlet Allocation or models provide a litmus test of how relations to machine learning are taking shape in the digital humanities. On the one hand, these models promise to make sense of large accumulations of documents (scientific publications, news, literature, online communications, etc.) in terms of underlying themes or latent "topics." As we will see, large document collections have long attracted the interest of machine learners. On the other hand, Jockers signals the practical difficulties of relating to machine learning when he suggests that "it is fair to skip the mathematics" for the sake of "the beauty of the result." Although some parts of the humanities and critical social research exhort closer attention to algorithms and mathematical abstractions, other parts elide its complexity in the name of "the beauty of the results."

Critical thought has not always endorsed the use of machine learning in digital humanities. Alex Galloway, for instance, makes two observations about the circulation of these methods in humanities scholarship. The first points to its marginal status in increasingly machine-learned media cultures:

> When using quantitative methodologies in the academy (spidering, sampling, surveying, parsing, and processing), one must compete broadly with the sorts of media enterprises at work in the contemporary technology sector. A cultural worker who deploys such methods is little more than a lesser Amazon or a lesser Equifax (Galloway 2014, 110).

Galloway highlights the asymmetry between humanities scholars and media enterprises or credit score agencies (Equifax). The "quantitative methodologies" that he refers to as spidering, sampling, processing, and so forth are more or less all epitomized in machine learning techniques (e.g., the Association Rule Mining techniques used by Amazon to recommend purchases or perhaps the decision tree techniques used by the credit-rating systems at Equifax and FICO (Fico 2015)). Galloway's argument is that the infrastructural scale of these enterprises, along with the sometime large technical workforces they employ to continually develop new predictive techniques, dwarfs any gain in efficacy that might accrue to humanities research in its recourse to such methods.

Galloway also observes that even if "cultural workers" do manage to learn to machine learn and become adept at repurposing the techniques in the interests of analyzing culture rather than selling things or generating credit scores, they might actually reinforce power asymmetries and exacerbate the ethical and political challenges posed by machine learning:

> Is it appropriate to deploy positivistic techniques against those self-same positivistic techniques? In a former time, such criticism would not have been valid or even necessary. Marx was writing against a system that laid no specific claims to the apparatus of knowledge production itself—even if it was fueled by a persistent and pernicious form of ideological misrecognition. Yet, today the state of affairs is entirely reversed. The new spirit of capitalism is found in

brainwork, self-measurement and self-fashioning, perpetual critique and innovation, data creation and extraction. In short, doing capitalist work and doing intellectual work—of any variety, bourgeois or progressive—are more aligned today than they have ever been (Galloway 2014, 110).

This perhaps more serious charge concerns the nature of any knowledge produced by machine learning. The "techniques" of machine learning may or may not be positivist, and indeed, given the claims that machine learning transforms the production of knowledge, positivism may not be any more stable than other conceptual abstractions. Hence, it might not be so strongly at odds with critical thought even if remains complicit—"aligned"—with capitalist work. Intellectual work of the kind associated with machine learning is definitely at the center of many governmental, media, business, and scientific fields of operation, and increasingly they anchor the operations of these fields. Neither observation—asymmetries in scale, alignment with a "positivist" capitalist knowledge economy—exhausts the potentials of machine learning, particularly if, as many people claim, it transforms the nature of knowledge production and hence "brainwork."

What cannot be automated?

Jaron Lanier's question—how will we conceive at a given time what cannot be automated?—suggests an alternative angle of approach. Like Galloway, I'm wary of certain deployments of machine learning, particularly the platform-based media empires and their efforts to capture sociality (Gillespie 2010; Van Dijck 2012). Machine learners do seem to be "laying claim to the apparatus of knowledge production." Yet even amid the jarring ephemera of targeted online advertising or the more elevated analytics of literary history, the transformations in knowledge and knowing do not automatically configure intellectual work as capitalist production. Empirical work to describe differences, negotiations, modifications, and contestation would be needed to show the unevenness, variability, and deep contingency of reconfigured knowledges. As I have already suggested, machine learning practice is not simply automating existing economic relations or even data practices. Although Hammerbacher and Galloway are understandably pessimistic about the existential gratifications and critical efficacy of building targeted advertising systems or document classifiers, the "deployment" of machine learning is not a finished process, but very much in train, constantly subject to revision, reconfiguration, and alteration.

Importantly, the familiar concerns of critical social thought to analyze differences, power, materiality, subject positions, agency, and so on somewhat overlap with the claims that machine learning produces knowledge of differences, of nature, cultural processes, communication, and conduct. Unlike other objects of critical thought, machine learners (understood always as human–machine ensembles) are closely interested in producing knowledge, albeit scientific, governmental, or operational. This coincidence of knowledge projects suggests the possibility of some different articulation, of modification of the practice of critical thought in its empirical and theoretical registers. The altered human–machine relations we see as machine learners might shift and be redrawn through experiments in empiricism and theory.

Where in the algorithms, calculations, abstractions, and regularizing practices of machine learning could differences be redrawn? Machine learning in journalism, in specific scientific fields, in the humanities, in social sciences, and in art, media, government, or civil society sometimes overflow the platform-based deployments and their trenchantly positivist usages. A fairly explicit awareness of the operation of machine learning-driven processes is taking shape in some quarters. This awareness supports a situationally aware calculative knowledge practice.

For instance, the campaign to reelect Barack Obama as U.S. president in 2011–2012 relied heavily on microtargeting of voters in the lead up to the election polls (Issenberg 2012; Mackenzie et al. 2016). In response to the data analytics-driven election campaign run by the U.S. Democrats, data journalists at the nonprofit news organization *ProPublica* reverse engineered the machine learning models that the Obama reelection team used to target individual votes with campaign messages (Larsen 2012). They built a machine learning model—the "Message Machine"—using emails sent in by voters to explore the workings of the Obama campaign team's microtargeting models. Although the algorithmic complexity and data infrastructures used in the Message Machine hardly match those at the disposal of the Obama team, it combines natural language processing (NLP) techniques such as measures of document similarity and machine learning models such as decision trees to disaggregate and map the microtargeting processes.

Reverse engineering work focused on the constitution of subject positions (the position of the "voter") can be found in other quarters. In response to the personalized viewing recommendations generated by streaming media service Netflix, journalists at *The Atlantic* working with Ian Bogost, a media theorist and programmer, reconstructed the algorithmic production of around 80,000 microgenres of cinema (Madrigal 2014). Although Netflix's system to categorize films relies on much manual classification and tagging with meta-data, the inordinate number of categories they use is typical of the classificatory regimes that are developing in machine learning-based settings.

Both cases address the contemporary configurations of doing, saying, and thinking subjects, not only to recognize how subject positions are assigned or made, but to

Table 1.1
A small sample of titles of scientific articles that use machine learning in relation to "difference"

Title	Year
A framework for measuring differences in data characteristics	2002
Computer-mediated knowledge sharing and individual user differences: an exploratory study	2004
Biclustering Gene Expression Data using MSR Difference Threshold	2009
Comparative Transcriptomic Approach To Investigate Differences in Wine Yeast Physiology and Metabolism during Fermentation	2009
Classifier-based analysis of visual inspection: Gender differences in decision-making	2010
SVM-Based Speaker Recognition Considering Gender Differences	2010
CHARACTERIZING MORPHOLOGY DIFFERENCES FROM IMAGE DATA USING A MODIFIED FISHER CRITERION	2011
Living Multiples: How Large-scale Scientific Data-mining Pursues Identity and Differences	2013
Differential Evolution with Temporal Difference Q-Learning Based Feature Selection for Motor Imagery EEG Data	2013
Constrained Least-Squares Density-Difference Estimation	2014

grasp the possibility of change. Although these cases may be exceptional achievements, and indeed highlight the dead weight of ad-tech application of machine learning, knowledge production more generally is not easily reducible to contemporary forms of capitalist labor.

Different fields in machine learning?

The proliferation of scientific machine learners suggests that the generalization of machine learning cannot be reduced to personalized advertising or other highly extractive uses. Table 1.1 presents a small sample of scientific literature at the intersection of "difference" and machine learning. This sample, although no doubt overshadowed by the fast-growing thicket of computer science publications on recommendation systems, targeted advertising or handwriting recognition, is typical of the positivity or specific forms of accumulation associated with machine learners in science. (I return to this topic in chapter 7 in discussing how the leveraging of scientific data via predictive models and classifiers deeply affects the fabric and composition of objects of scientific knowledge.) The longevity and plurality of experiments, variants, alternative techniques, implementations, and understandings associated with machine learning

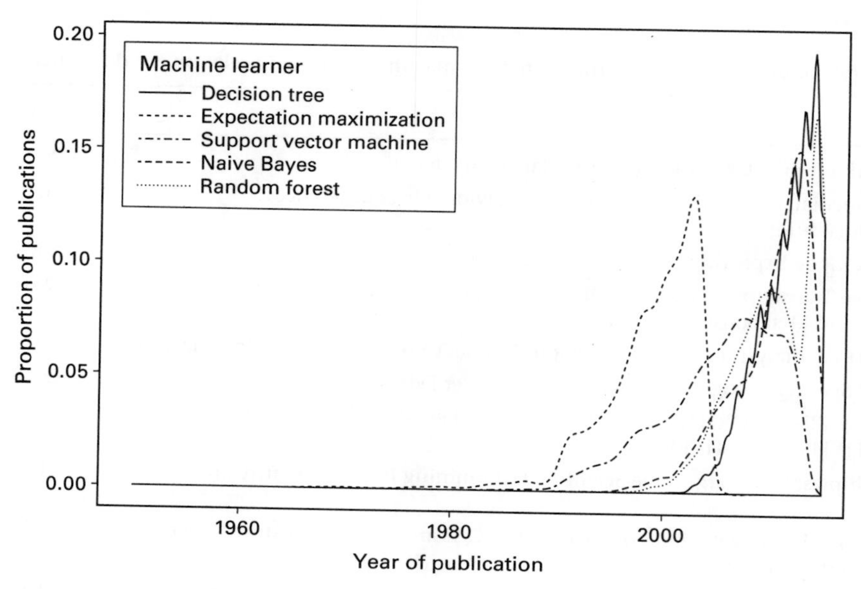

Figure 1.3

Machine learners in scientific literature. The lines in the graph suggest something of the changing fortunes of machine learners over time. The publication data come from Thomson Reuter's *Web of Science*. Separate searches were run for each machine learner. In these plots, as in the Google Trends data, the actual counts of publications have been normalized. In contrast to the Google Trends plots, these plots do not show the relative counts of the publications, only their distribution in time.

makes it difficult to immediately reduce them to capitalist captures of knowledge production.

If we attend to the flow of machine learning practices, devices, and techniques in scientific fields, then a diversification rather than a simple scaling-up to industrial-strength infrastructures begins to appear. Figure 1.3 derives from counts of scientific publications that mention particular machine learners such as `decision tree` or `Naive Bayes` in their title, abstract, or keywords. The curves, which are probability density plots, suggest a time-varying distribution of statements and operations for different techniques. This crude plot outlines the duration and the ebbs and flows of work on specific techniques, platforms, knowledges, and power relations. Like the Google Trends searches for `machine learning`, the lines shown in figure 1.3 have been normalized to adjust for an overall increase in the volume of scientific publications over the last five decades. Unlike the Google Trends search patterns, the scientific literature displays polymorphous temporalities in which different techniques and operations diverge widely from each other over the last half century. Crucially for my purposes, machine learning in the sciences constitutes an a-totality, a heterogeneous volume and decentered production of statements.

The diagram in critical thought

An experiment in the practice of critical thought amid the accumulations of data, settings, and devices, this book archaeologically diagrams the data practices of machine learning with respect to knowledge production. Despite their many operational deployments, the coming together of algorithm, calculation, and technique in machine learning is not fully coherent or complete. To qualify or specify how machine learners exist in their generality, we would need to specify their operations at a level of abstraction that neither attributes a mathematical or algorithmic ideality to them nor frames them as yet another means of production of relative surplus value. Finding ways of articulating their diversity, loose couplings, and mutability would assist in understanding their operational power and orient us to their potential to create new forms of difference.[5]

The experiment will rely on diagrams. Diagrams—a form of drawing that smooths away many frictions and variations—practically abstract. (Gilles Deleuze, in his account of Michel Foucault's philosophy, presents diagrams as centers of power knowledge: "What is a diagram? It is a display of the relations between forces which constitute power in the above conditions... the diagram or abstract machine is the map of relations between forces, a map of destiny, or intensity" (Deleuze 1988b, 36)). Diagrams retain a connection to forms of doing, such as "learning from experience," that

5 Certain strands of social and cultural theory have taken a strong interest in algorithmic processes as operational forms of power. For instance, the sociologist Scott Lash distinguishes the operational rules found in algorithms from the regulative and constitutive rules studied by many social scientists:

> in a society of pervasive media and ubiquitous coding, at stake is a third type of rule, algorithmic, generative rules. "Generative" rules are, as it were, virtuals that generate a whole variety of actuals. They are compressed and hidden and we do not encounter them in the way that we encounter constitutive and regulative rules. Yet this third type of generative rules is more and more pervasive in our social and cultural life of the post-hegemonic order. They do not merely open up opportunity for invention, however. They are also pathways through which capitalist power works, in, for example, biotechnology companies and software giants more generally (Lash 2007, 71).

The term "generative" is somewhat resonant in the field of machine learning as generative models, models that treat modeling as a problem of specifying the operations or dynamics that could have given rise to the observed data, are extremely important.

idea-centric accounts of abstraction sometimes struggle with. Perceptually and operationally, they span and indeed criss-cross between human and machine–machine learners. As we will see, diagrams form an axis of human–machine encounters in machine learning and a significant site of invention and emergence. They accommodate compositional substitutions, variations, and superimpositions, as well as a play or movement among their often heterogeneous elements. Occasionally, perhaps rarely (Foucault as we will see characterizes statements by their rarity amid accumulation), by virtue of its composition, diagrams bring something new into the world.

Both the commonality and specificity (indeed the specific commonality) of machine learners would be hard to grasp without being able to trace their diagrammatic composition. Similarly, to understand the operational formation associated with machine learning, the connections among data structures, infrastructures, processors, databases, and lives need to be mapped. One path to follow in doing this—certainly not the one for everyone—is to partially reconfigure oneself as a machine learner by occupying operational subject positions (the programmer or software developer, the statistics or computer science student, the modeler, the researcher, the data scientist, etc.). In moving among some of these subject positions, the densely operational indexes of mathematical formalisms begin to unravel.

Diagrams can be drawn in multiple ways using various materials and inscriptive practices. At times unwittingly and naively interpelated by the claim on machine learners to know differently, I use code and software implementations, graphical plots, and mathematical expressions absorbed or copied from textbooks, blogs, online videos, and the heavy accumulation of scientific publications from many disciplines (e.g., as seen in figure 1.3), together with the theoretical resources of a media-focused archaeology of knowledge and a science studies-informed ethnographic sensibility toward always situated and configured infrastructural and calculative practices.

From my own learning to machine learn, I draw six major machine learning operations diagrammatically: vectorization, optimization, probabilization, pattern recognition, regularization, and propagation. These generic operations intersect in a diagram of machine learning spanning hardware and software architectures, organizations of data and datasets, practices of designing and testing models, intersections between scientific and engineering disciplines, and professional and popular pedagogies. With varying degrees of formalization and consistency, these operations might also occasion or provoke some creative, resistive, or redistributive moves for critical thought in relation to differences, materiality, experience, agency, or power. A summary of the operations can also be found near the beginning of the concluding chapter.

My somewhat risky, naive immersion in technical practice seeks to support an alternative account of machine learning, an account in which some schema, analogy, imagining, and sense of agency can take root. Mundane technical practices, sometimes at a quite low level (e.g., vectorization) and other times at a high level of formalization (e.g., in discussing mathematical functions), are elements to be drawn—sometimes literally, sometimes operationally—on a diagram. The archaeology of the operational formation of machine learning does not unearth the foundations of a strategic monolith but graphs the local striations of force that feed into the generalization and plurality of the field in both its monumental accumulations and peripheral mutations.

2 Diagramming Machines

Machine learning is not magic; it cannot get something from nothing. What it does is get more from less. Programming, like all engineering, is a lot of work: we have to build everything from scratch. Learning is more like farming, which lets nature do most of the work. Farmers combine seeds with nutrients to grow crops. Learners combine knowledge with data to grow programs (Domingos 2012, 81).

The tools or material machines have to be chosen first of all by a diagram (Deleuze 1988b, 39).

The "learning" in machine learning embodies a change in programming practice or indeed the programmability of machines. Our sense of the potentials of machine learning can be understood, according to Pedro Domingos, in terms of a contrast between programming as "a lot of [building] work" and the "farming" done by machine learners to "grow programs." In characterizing machine learning, the tensions between the programming "we" (programmers, computer scientists?) do and the programming that learners do ("growing") are worth pursuing. Although Domingos suggests that machine learners "get more from less," I will propose that an immense constellation of documents, software, publications, blog pages, books, spreadsheets, databases, data center architectures, whiteboard and blackboard drawings, and an inordinate amount of talk and visual media feed into the growth of machine learners. Lively growth has occurred in machine learning, but this liveliness and the sometimes life-like growth of machine learners are a regional expression of a distributed formation. "Wherever there is a region of nature," the philosopher Alfred North Whitehead writes, "which is itself the primary field of the expressions issuing from each of its parts, that region is alive" (Whitehead 1956, 31).

In this chapter, I identify and describe some distributed practices that give rise to a sense of machine learners as growing or lively. I will argue that these practices, issuing from many parts, can be traced partially through code written in generic or specialized programming languages such as `Python` and `R`, in libraries of machine learning code such as `R`'s `caret` or `Python`'s `scikit-learn`, or `TensorFlow` to do machine learning.

Code obscures and reveals multiple transformations at work in the operational formation. Science studies scholars such as Anne-Marie Mol have affirmed the need to keep practice together with theories of what exists. Mol writes:

> If it is not removed from the practices that sustain it, reality is multiple. This may be read as a description that beautifully fits the facts. But attending to the multiplicity of reality is also an act. It is something that may be done—or left undone (Mol 2003, 6).

Mol advocates thinking of reality as practically multiple. Her insistence on the nexus of practice or doing and the plural existence of things suggests a way of handling the code that machine learners produce. Code should be approached in its multiplicity. In the case of machine learners, this means weaving pedagogical expositions of machine learning focused on mathematical derivations into the accumulation of scientific or technical research publications, ranging from textbooks to research articles, which vary, explore, experiment, and implement machine learners in code. Reading and writing code alongside scientific papers, YouTube lectures, machine learning books, and competitions is not only a form of observant participation but directly forms part of the diagrammatic multiplicity under description.

Although machine learning utterly depends on code, I will suggest that no matter how expressive or well documented it may be, code alone cannot fully diagram how machine learners make programs or how they combine knowledge with data. Domingos writes that "learning algorithms . . . are algorithms that make other algorithms. With machine learning, computers write their own programs, so we don't have to" (Domingos 2015, 6). Yet the writing performed by machine learners cannot be read textually or procedurally as programs might be read, for instance, in work known as critical code studies. The difference between the code of an Atari computer game console in Nick Montfort and Ian Bogost's *Racing the Beam* (Montfort and Bogost 2009) and the machine learning of Atari's games undertaken by DeepMind in London during recent years (Mnih et al. 2013, 2015) is hard to access in program code. The learning or making by learning is far from homogenous, stable, or automatic in practice. Materially, code is only one element in the diagram of machine learning. It displays, with greater or lesser degrees of visibility, relations among a variety of forces (infrastructures, scientific knowledges, mathematical formalisations, etc.). It is aligned by and exposes multiple institutional, infrastructural, epistemic, and economic positions.

Coding practices and the pedagogical expositions of machine learning have shifted substantially over the last decade or so due to the growth in open source programming languages and as part of the broader and well-known expansion of digital media cultures. The fact that data scientists, software developers, and other machine learners across scientific and commercial settings use programming languages such as `Python` and `R` more than specialized commercial statistical and data software packages such as `Matlab`, `SAS`, or `SPSS` (Muenchen 2014) is perhaps symptomatic of shifts in computational culture. Coding cultures are crucial to the recent growth of machine learning. Although scientific computing languages such as `FORTRAN`—"Formula Translator"—have long underpinned scientific research and engineering applications in various fields (Campbell-Kelly 2003, 34–35), the development in recent decades of data-analytic and statistical programming languages and coding frameworks has crystallized a repertoire of standard operations, patterns, and functions for reshaping data and constructing models that classify and predict events and associations among things, people, processes, and so on. This development continues apace, especially in research and engineering driven by social media and internet platforms such as Facebook and Google. Although Domingos speaks of "growing" programs, the accumulating sediment of coding and related data practices are the soil in which machine learners take root. The different elements of coding practice are the faceted attachments and associations that we need to access and traverse to know and come to grips empirically with contemporary compositions of power and knowledge in machine learning.

"We don't have to write programs"?

In machine learning, coding changes from what we might call symbolic logical diagrams to statistical algorithmic diagrams. Although many machine learning techniques have long statistical lineages (running back to the 1900s in the case of Karl Pearson's development of the still-heavily used Principal Component Analysis (Pearson 1901)), machine learners often distance themselves from the classical computer science understanding of programs as manipulation of symbols, even as they rely on such symbolic operations to function. Symbolic manipulation, epitomized by deductive logic or predicate calculus, was very much at the center of many artificial intelligence (AI) projects during the 1950s and 1960s (Dreyfus 1972; Edwards 1996). In machine learning, the privileged symbolic-cognitive forms of logic are subject to a statistical transformation.

Take, for instance, one of the most common operations of the Boolean logical calculus, the `NOT-AND` or `NAND` function shown in table 2.1. The truth table summarizes a logical function that combines three input variables `X1`, `X2`, and `X3` and produces the output variable `Y`. Because in Boolean calculus, variables or predicates can only take the values `true` or `false`, they can be coded in as `1` and `0`.

Now in Foucauldian terms, the truth table and its component propositions constitute a *statement*. This statement has triple relevance for archaeology of machine

Table 2.1
The truth table for the Boolean function NOT-AND truth

X1	X2	X3	Y
0	0	0	1
0	0	1	1
0	1	1	1
0	1	0	1
1	0	0	1
1	0	1	1
1	1	0	1
1	1	1	0

learning. The spatial arrangement of the table is fundamental to the ambitions of machine learners (and this is the topic of chapter 3). Most datasets come as tables or end up as tables at some point in their analysis. Second, the elements or cells of this table are numbers. The numbers 1 and 0 are the binary digits as well as the "truth" values "True" and "False" in classical logic. These numbers are readable as symbolic logical propositions governed by the rule $Y = \neg X_1 \wedge X_2 \wedge X_3$. The table acts as a hinge between numbers and symbolic thought or cognition. Third, the NAND table in particular has an obvious operational relevance in digital logic because digital circuits of all kinds—memory, processing, and communication—comprise such logical functions knitted together in the intricate gated labyrinths of contemporary calculation.[1]

A pre-machine learning programmer, tasked with implementing the logical NAND function, might write:

Listing 2.1
XOR function in Python

```
Y = not(X1 & X2 & X3)
```

The trivial simplicity of the code stands out. This looks like the kind of symbol manipulation that computers can easily be programmed to do. How, by contrast, could such a truth table be learned by a machine, even a machine whose *modus operandi*

1 The philosophy Charles Sanders Peirce had first shown that combinations of NAND operations could stand in for any logical expression whatsoever, thus paving the way for the diagrammatic weave of contemporary digital memory and computation in all their permutations. Today, NAND-based logic is norm in digital electronics. NOR—NOT OR—logic is also used in certain applications.

and indeed whose very fabric is nothing other than a massive mosaic of NAND operations inscribed in semiconductors? Machine learning of on elementary truth tables is not a typical operation today, but usefully illustrates something of the diagrammatic transformations that programming or coding (and classical AI) has undergone.

Machine learners such as decision trees or neural nets typically know nothing of the logical calculus and its rules. Can they be induced to learn them? A perceptron, an elementary machine learner dating from the 1950s, that "learns" the binary logical operation NAND (Not-AND) is expressed in 20 lines of python code on the Wikipedia "Perceptron" page (*Perceptron* 2013) (see listing 2.2). It is a standard machine learner almost always included in machine learning textbooks and usually taught in introductory machine learning classes.

Listing 2.2
Perceptron learning *XOR* function

```
threshold = 0.5
learning_rate = 0.1
weights = [0, 0, 0]
training_set = [((1, 0, 0), 1), ((1, 0, 1), 1), ((1, 1, 0), 1),
    ↪ ((1, 1, 1), 0)]

def dot_product(values):
    return sum(value * weight for value, weight in zip(values,
        ↪ weights))

out = 'weight1, weight2, weight3, error_count, iteration'
print  out
iteration_count = 0
while True:
    error_count = 0
    iteration_count += 1
    for input_vector, desired_output in training_set:
        out = ','.join(map(str,weights)) + ',' + str(error_count)
        out = out + ',' + str(iteration_count)
        result = dot_product(input_vector) > threshold
        error = desired_output - result
        print out
        if error != 0:
            error_count += 1
            for index, value in enumerate(input_vector):
                weights[index] += learning_rate * error * value
    if error_count == 0:
        break
```

Table 2.2
Iterative change in weights as a perceptron learns the NAND function

weight1	weight2	weight3	error_count	iteration
0.00	0.00	0.00	0	1
0.10	0.00	0.00	1	1
0.20	0.00	0.10	2	1
0.30	0.10	0.10	3	1
0.30	0.10	0.10	0	2
0.40	0.10	0.10	1	2
0.50	0.10	0.20	2	2
0.50	0.10	0.20	2	2
0.40	0.00	0.10	0	3
0.50	0.00	0.10	1	3
0.50	0.00	0.10	1	3
0.60	0.10	0.10	2	3
0.50	0.00	0.00	0	4
0.60	0.00	0.00	1	4
0.60	0.00	0.00	1	4

What does the code in listing 2.2 show or say about the transformation in programmability or the writing of programs? First, we should note the relative conciseness of the code vignette. Much of the code here is familiar, generic programming. It defines variables, sets up data structures (lists of numerical values), checks conditions, loops through statements, or prints results. In citing this code, I am not resorting to a technical publication or scientific literature as such or even to a machine learning software library or package (in `scikit-learn`, the same model could be written `p = scikit-learn.linearmode.Perceptron(X,Y)`), just to a Wikipedia page and the relatively generic and widely used programming language `Python`.[^1.31] Whatever the levels of abstraction associated with machine learning, the code is hardly ever hermetically opaque. As statements, everything lies on the surface.

Second, although the shaping of data, the counting of errors, and the optimization of models are topics of later discussion, the code shows some typical features of a machine learner in the form of elements such the `learning rate`, a `training_set`, `weights`, an `error count`, and a loop function that multiplies values (`dot_product`). Some of the names, such as `learning_rate` or `error_count` present in the code, bear the marks of the theory of learning machines that we will discuss.

Third, executing this code (e.g., by copying and pasting it into a `Python` terminal) produces several dozen lines of numbers. They are initially different from each other but gradually converge on the same values (see table 2.2). These numbers are the "weights" of the nodes of the perceptron as it iteratively "learns" to recognize patterns in the input values. None of the workings of the perceptron need concern us at the moment. Again, what runs across all of these observations are the numbers that the algorithm produces as output—they embody the program that the perceptron has written. How has the learning happened in code? The NAND truth table has been redrawn as a dataset (see line 4 of the code that defines the variable `training_set`). The perceptron has learned the data by approaching it as a set of training examples and then adjusting its internal model–the weights that are printed during each loop of the model as the output–repeatedly until the model is producing the correct result values *Y* of the truth table. The algorithm exits its main loop (`while True:`) when there are no errors.

The perceptron algorithm computes numbers—`0.79999`, `2.0`—as weights or parameters. These numbers display no direct correspondence with the symbolic categories of Boolean true and false or the binary digits `1` and `0`. There may be a relation, but it is not obvious at first glance. The problem of mapping these calculated parameters—and they truly abound in machine learning—triggers many different diagrammatic movements in machine learning (and these different movements will be discussed in chapters 3 and 4). These numbers engender much statistical ratiocination (see chapter 5). Here we need only note the contrast between symbolically organized statements like the NAND truth table of table 2.1 and the operational statements in table 2.2.

The operation here is recursive: a model or algorithm implemented in digital logic (as `Python` code) has "learned"—a term we need to explore more carefully—a basic rule of digital logic (the `NAND` function) by treating logical propositions as data. This mode of transformation is symptomatic. The learning done in machine learning has few cognitive or symbolic underpinnings. It differs from classical AI in that it treats existing symbolic, control, communicative, and, increasingly, signifying processes (such as the cat faces that `kittydar` tries to find) and latches onto them programmatically only in the form of weights.

The elements of machine learning

If this growing of programs through modeling data is a different mode of coding operations and a different mode of knowing, how does one learn to do it? In the course of writing this book, as well as reading the academic textbooks, popular how-to books, software manuals, help documents, and blog-how-to posts, I attended graduate courses in Bayesian statistics, genomic data analysis, data mining, and missing data. I also participated in online machine learning courses and some machine learning competitions. These are all widely shared activities for people learning to do machine learning. I watched and copied down by hand many statements, equations, and propositions from

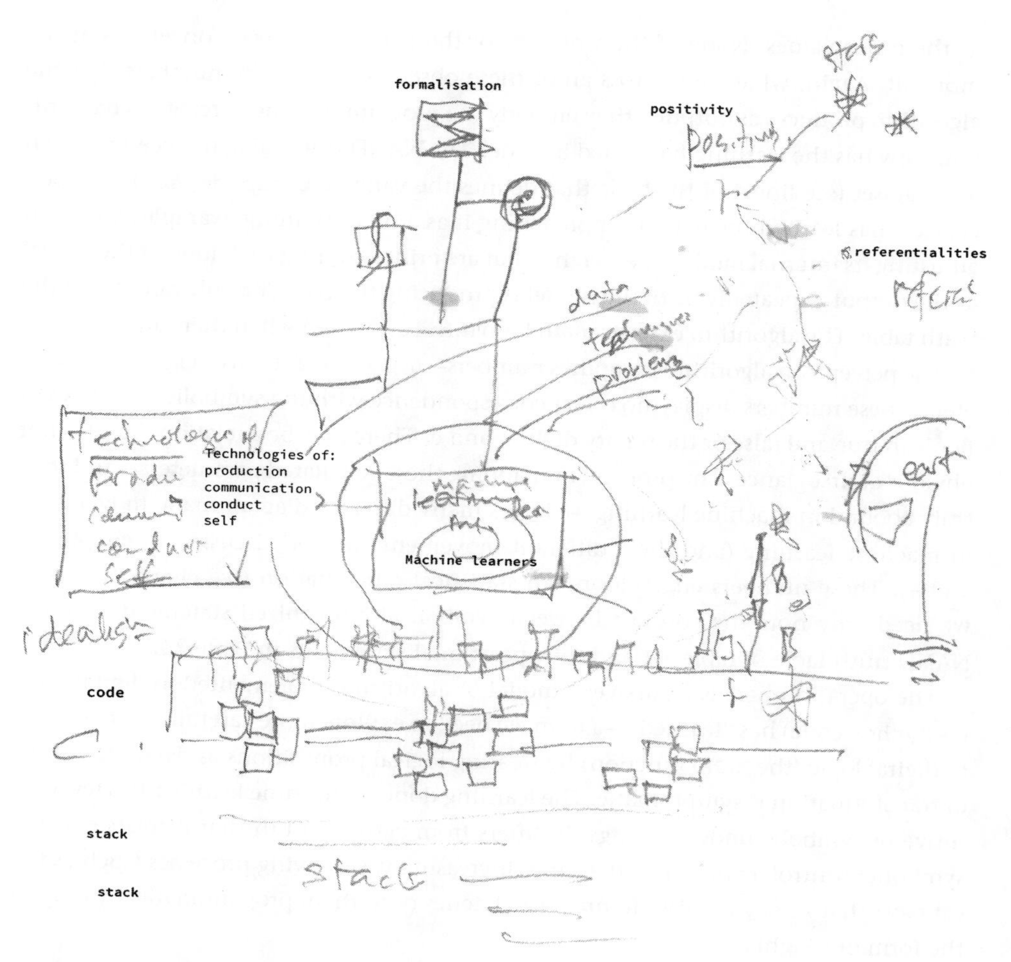

Figure 2.1
Elements of learning to machine learn.

YouTube videos of the 18 lectures in Andrew Ng's Stanford University lectures CS229 computer science course recorded in 2008. These lectures have cumulative viewing figures of around 500,000 (Ng 2008a). I spent extended hours learning to use various code libraries, packages, and platforms in `R`, `Python`, and, to a limited extent, `Torch`. Finally, I gradually accumulated and worked with a set of around 400,000 articles drawn from various fields of science in response to search queries on particular machine learners, such as `support vector machine` or `expectation maximization`. This accumulation of knowledge appears in figure 2.1 as a network-cloud.

Some broadly shared topic structures help in navigating the software libraries, pedagogical, and research literatures. The textbooks, the how-to recipe books (Segaran 2007; Kirk 2014; Russell 2011; Conway and White 2012), and the online university courses on machine learning often have a similar topic structure. They nearly always begin with "linear models" (fitting a line to the data), then move to logistic regression (a way of using the linear model to classify binary outcomes; e.g., spam/not-spam; malignant/benign; cat/non-cat), and afterward move to some selection of neural networks, decision trees, support vector machine, and clustering algorithms. They add in some decision theory, techniques of optimization, and ways of selecting predictive models (especially the bias-variance trade-off).[2] The topic structures have in recent years started to become increasingly uniform. This coagulation around certain topics,

2 They differ, however, in several important respects. Reading the *Elements of Statistical Learning* textbook or one of the machine learning books written for programmers (e.g., *Programming Collective Intelligence* or *Machine Learning for Hackers* (Segaran 2007; Conway and White 2012)) does not directly subject the reader to machine learning. By contrast, doing a Coursera course on machine learning brings with it an ineluctable sense of being machine-learned, of oneself becoming an object of machine learning. The students on Coursera are the target of machine learning. Daphne Koller and Andrew Ng are leading researchers in the field of machine learning, but they also cofounded the online learning site Coursera. As experts in machine learning, it is hard to imagine how they would not treat teaching as a learning problem. Indeed, Daphne Koller sees things this way:

> There are some tremendous opportunities to be had from this kind of framework. The first is that it has the potential of giving us a completely unprecedented look into understanding human learning. Because the data that we can collect here is unique. You can collect every click, every homework submission, every forum post from tens of thousands of students. So you can turn the study of human learning from the hypothesis-driven mode to the data-driven mode, a transformation that, for example, has revolutionized biology. You can use these data to understand fundamental questions like, what are good learning strategies that are effective versus ones that are not? And in the context of particular courses, you can ask questions like, what are some of the misconceptions that are more common and how do we help students fix them? (Koller 2012)

Whether the turn from "hypothesis-driven mode to the data-driven mode" has "revolutionized biology" is debatable (I return to this in a later chapter). Whether the data generated by my participation in Coursera's courses on machine learning generate data supports understanding of fundamental questions about learning also seems an open question. Nevertheless, the loopiness of this description interests and appeals to me. I learn about machine learning, a way for computer models to optimize their predictions on the basis of "experience"/data, but at the same time, my learning is learned by machine learners. This could not happen easily with a printed text, although versions of it happen all the time as teachers work with students on reading texts. Although Coursera and other MOOCs promise something that mass education struggles to offer (individually profiled educational services), it also negatively highlights the possibility that machine learning in practice can, somewhat recursively, help us make sense of machine learning as it develops.

problems, and mathematical formalisms is something worth analyzing (e.g., because it definitely affects how machine learning is taken up in different settings), but it should not be taken as obvious or given because it results from many iterations.

Amid this avalanching machine learning materials and practice, a single highly cited and compendious textbook, *Elements of Statistical Learning: Data Mining, Inference, and Prediction* dating from from around 2000 and currently in its second edition (Hastie, Tibshirani, and Friedman 2009), can be seen from almost any point of the terrain.[3] At least for archaeological purposes, I regard this book as an assemblage and a diagram that presents many important statements, forms of visibility, and relations between forces at work in machine learning. The authors of the book, Jeff Hastie, Rob Tibshirani, and Jerome Friedman, are statisticians working at Stanford and Columbia University.

Elements of Statistical Learning is a massive textual object, densely radiant with equations, tables, algorithms, graphs, and references to other scientific literature. From the first pages proper of the book, almost every page has a figure, table, or formal algorithm (counting these together: 1670 equations, 291 figures, 34 tables, and 94 algorithms, for a total of 2089 operational statements threaded through the book). Equations rivet the text with mathematical abstractions of varying sophistication. On each page of the book, we see reading, puzzling over, and perhaps learning from the products of code execution. The graphic figures are all produced by code. The tables are mostly produced by code. The algorithms specify how to implement code, and the equations diagram various operations, spaces, and movements that meant to run as code.

In the range of references, combinations of code, diagram, equation, scientific disciplines, and computational elements, and perhaps in the somewhat viscous, interobjectively diverse referentiality that impinges on any reading of it, *Elements of Statistical Learning* betrays some hyperobject-like positivity (Morton 2013). It is an accumulation of forms, techniques, practices, propositions, and referential relations. *Elements of Statistical Learning* combines statistical science with various algorithms to "learn from data" (Hastie, Tibshirani, and Friedman 2009, 1). The data range across various kinds of problems (identifying spam email, predicting risk of heart disease, recognising handwritten digits, etc.). The learning takes the form of various machine learning techniques, methods, and algorithms (linear regression, k-nearest neighbors, neural networks, support vector machines, the Google Page Rank algorithm, etc.).

3 The complete text of the book can be downloaded from the website http://statweb.stanford.edu/~tibs/ElemStatLearn/. At the end of short intensive course on data mining at the Centre of Postgraduate Statistics, Lancaster University, the course convenor, Brian Francis, recommended this book as the authoritative text. I took that recommendation seriously. That book, although valuable in many respects, bristles with distracting formalizations and tacit practices.

There are other juggernaut machine learning textbooks. Ethem Alpaydin's *Introduction to Machine Learning* (Alpaydin 2010) (a more computer science-based account), Christopher Bishop's heavily mathematical *Pattern Recognition and Machine Learning* (Bishop 2006), Brian Ripley's luminously illustrated and almost coffee table-formatted *Pattern Recognition and Neural Networks* (Ripley 1996), Tom Mitchell's earlier artificial intelligence-centered *Machine Learning* (Mitchell 1997), Peter Flach's perspicuous *Machine Learning: The Art and Science of Algorithms That Make Sense of Data* (Flach 2012), or, further afield, the sobering and laconic *Statistical Learning for Biomedical Data* (Malley, Malley, and Pajevic 2011) all cover a similar range of data and approaches. These and quite a few other recent machine learning textbooks display a range of emphases, ranging from the highly theoretical to the practical, from an orientation to statistical inference to an emphasis on computational processes, from science to commercial applications.

Who reads machine learning textbooks?

How do machine learning textbooks help us assess and engage with the claim to learn from data or to produce knowledge differently? Although *Elements of Statistical Learning* certainly does not comprehend everything taking place in and around machine learning, it diagrams several *elementary* tendencies or traits. Its readership, as we will see, is widespread. It has a heterogeneous texture in terms of the examples, formalisms, disciplines, and domains it covers. It starkly renders the problems of making sense of mathematical operations, diagrams, and transformations carried on through calculation, simulation, deduction, or analysis. It draws on a matrix of operational practices, particularly in the form of the `R` code it heavily but somewhat latently relies on. In short, *Elements of Statistical Learning* presents a multifaceted and somewhat monumental montage of abstractive practice that might be open to archaeological inquiry.

Who reads the *Elements of Statistical Learning*? It is often cited by academic machine learning practitioners as an authoritative guide. However, students participating in new data science courses often come from different disciplinary backgrounds and find the tome unhelpful (see the comment by students during an introductory data science course documented in (Schutt and O'Neil 2013)). Regardless of whether the citations are friendly, it is hard to find a field of contemporary science, engineering, natural, applied, health, and indeed social science that has not cited it. A Thomson-Reuters Scientific "Web of Science" (TM) search for references citing either the first or second edition of (Hastie, Tibshirani, and Friedman 2009) yields around 9000 results. These publications sprawl across more than 100 different fields of research. Although

Table 2.3
Subject categories of research publications citing *Elements of Statistical Learning* 2001–2015. The subject categories derive from Thomson-Reuter *Web of Science*.

Citations	Field
556	Engineering
520	Computer Science
380	Engineering", "Computer Science
369	Biotechnology & Applied Microbiology
344	Mathematical & Computational Biology
211	Mathematics", "Computer Science
175	Chemistry
150	Remote Sensing
116	Instruments & Instrumentation
97	Imaging Science & Photographic Technology", "Computer Science
78	Engineering", "Automation & Control Systems
77	Public, Environmental & Occupational Health
75	Medical Informatics
75	Operations Research & Management Science", "Computer Science
74	Mathematics", "Biochemistry & Molecular Biology
69	Mathematics
67	Research & Experimental Medicine
66	Geology
62	Marine & Freshwater Biology
59	Engineering", "Engineering

computer science, mathematics, and statistics dominate, a diverse set of references comes from disciplines from archaeology, through fisheries and forestry, genetics, robotics, telecommunications, and toxicology ripple out from this book since 2001. Table 2.3 shows the top 20 fields by count. One could learn something about the diagrammatic movement of machine learners from that reference list, which spans biomedical, engineering, telecommunications, ecology, operations research, and many other fields. Although it is not surprising to see computer science, mathematics, and engineering appearing at highest concentration in the literature, molecular biology, control and automation, operation research, business, and public health soon appear, suggesting something of the propagating accumulation or positivity of machine learning.

So we know that *Elements of Statistical Learning* passes into many fields. But what do people read in the textual environment of book? In general, the thousands of citations of the book compose a diagram of the book's intersection with different domains of knowledge in operation. The relative concentrations and sparsities of these citations suggest there may be specific sites of engagement in the techniques, approaches, and machines that the book documents. Of the around 760 pages in the two editions

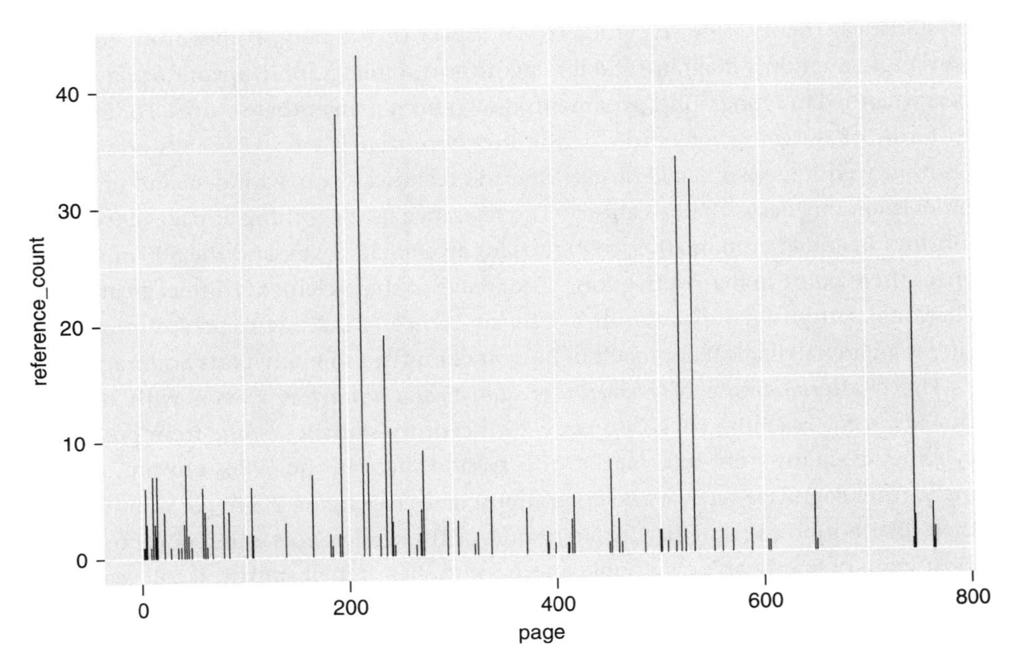

Figure 2.2
Pages cited from *Elements of Statistical Learning* by academic publications in all fields.

((Hastie, Tibshirani, and Friedman 2009) and (Hastie, Tibshirani, and Friedman 2001)), around 77 distinct pages are referenced in the citing literature. As figure 2.2 indicates, certain portions of the book are much more heavily cited than others. This distribution of page references in the literature that cites *Elements of Statistical Learning* is a rough guide to how the book has transited different settings. For instance, the most commonly cited page in the book is page 553. That page begins a section called "Non-negative Matrix Factorization," a technique frequently used to process digital images to compress their visual complexity into a simpler set of visual signals (Hastie, Tibshirani, and Friedman 2009, 553). (The underlying reference here is the highly cited paper (Lee and Seung 1999).) Like `kittydar`, it, as Hastie and co-authors write, "learns to represent faces with a set of basis images resembling parts of faces" (Hastie, Tibshirani, and Friedman 2009, 555). (So `kittydar`, which doesn't use NMF, might do better if it did because it could work just with parts of the images that lie somewhere near the parts of a cat's face—its nose, its eyes, its ears.)

Conversely, what do the authors of *Elements of Statistical Learning* read? The book gathers elements from many different quarters and seeks to integrate them in terms of statistical theory. The hyperobject-like aspect of the book comes from the thick weave of equations, diagrams, tables, algorithms, bibliographic apparatus, and numbers wreathed in typographic ornaments drawn from many other sources. For instance, in terms of outgoing references or the literature that it cites, *Elements of Statistical Learning* webs together a field of scientific and technical work with data and predictive models ranging across half a century. The reference list beginning at page 699 (Hastie, Tibshirani, and Friedman 2009, 699) runs for around 35 pages, and the 500 or so references there point in many directions. The weave of these elements differs greatly from citational patterns in the humanities or social sciences. Reading this book almost necessitates an archaeological approach because it comprises so many parts and fragments.

The citational fabric of *Elements of Statistical Learning* is woven with different threads, some reaching back into early 20th-century statistics, some from post-WW2 cybernetics, many from information theory, and then, in the 1980s onward, increasingly, from cognitive science and computer science (see figure 2.3 to get some sense of their distribution over time). Although some of these references either point to Hastie, Tibshirani, or Friedman's own publications, or to that of their statistical colleagues, the references rove quite widely in other fields and over time. *Elements of Statistical Learning* as a text is processing, assimilating, and recombining techniques, diagrams, and data from many different places and times. (The different waves appearing in the references cited in (Hastie, Tibshirani, and Friedman 2009) will shape discussion in later chapters in certain ways; for instance, late 1990s biology is the topic of chapter 7 and optimization functions dating from the 1950s are discussed in chapter 4.)

Both the inward and outward movements of citation suggest that *Elements of Statistical Learning*, like much in the field of machine learning, has a matrix-like character that constantly superimposes and transposes elements across boundaries and barriers. The implication here is that machine learning as a knowledge practice has a highly interwoven texture and in this respect differs somewhat from the classical understandings of scientific disciplines as bounded by communities of practice, norms, and problems (e.g., as in Thomas Kuhn's account of normal science (Kuhn 1996)). This aggregate or superimposed character of machine learning should definitely figure in any sense we make of it and will affect how critical thought addresses machine learners.

R: A matrix of transformations

Although barely a single line of code appears in *Elements of Statistical Learning*, all of the learners presented there are implemented in a single programming language, R. Coding is the operational practice that links the different planes and elements of machine

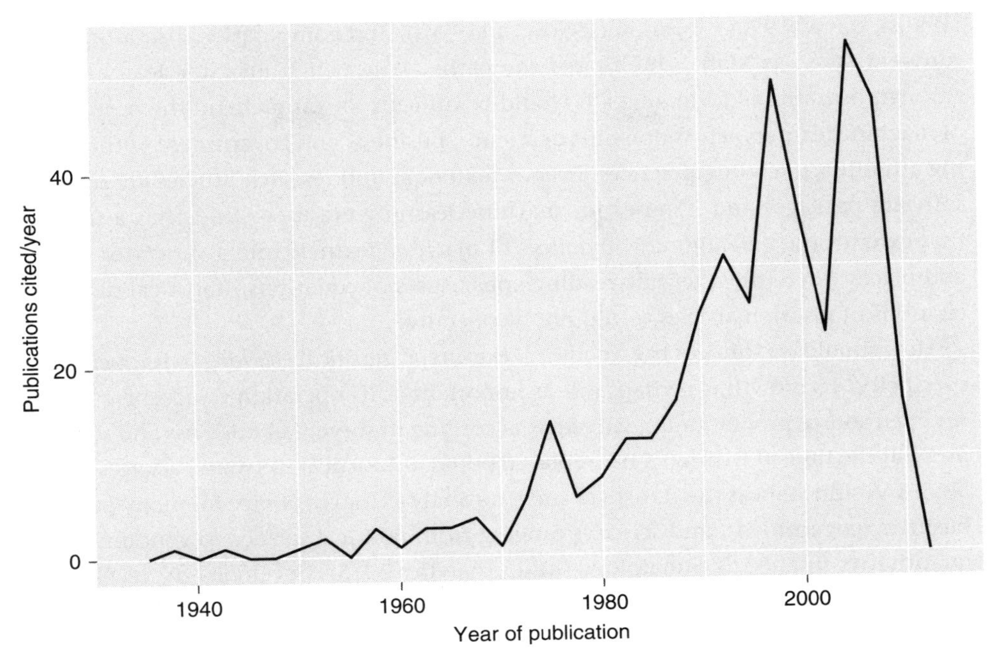

Figure 2.3
Publications cited in *Elements of Statistical Learning*. The references range over almost 80 years, with peaks in late 1970s, late 1980s, mid-1990s, and mid-2000s. These peaks relate to different mixtures of cybernetics, statistics, computer science, medicine, biology, and other fields running through machine learning. Regression-related publications form the main body of citation.

learning in an operational formation. The authors mention that "we used the R and S-PLUS programming languages in our courses" (Hastie, Tibshirani, and Friedman 2009, 9), but many elements of the book derive from `R` code.[4] The proliferation of programming languages such as `FORTRAN` (dating from the 1950s), `C` (1970s), `C++` (1980s), `Perl`

4 In a later book by some of the same authors with the similar sounding title, *An Introduction to Statistical Machine Learning with Applications in* `R` (James et al. 2013), `R` does appear in abundance. This book, however, is much shorter and less inclusive in various ways. There are in any case many online manuals, guides, and tutorials relating to `R` (Wikibooks 2013). For present purposes, I draw mainly on semi-popular books such as *R in a Nutshell* (Adler and Beyer 2010), *The Art of* `R` *Programming* (Matloff 2011), *R Cookbook* (Teetor 2011), *Machine Learning with R* (Lantz 2013), or *An Introduction to Statistical Learning with Applications in R* (James et al. 2013). These books are not written for academic audiences, although academics often write and use them in their work. They are largely made up of illustrations and examples of how to do different things with different types of data, and their examples are typically oriented toward business or commercial settings,

(1990s), Java (1990s), Python (1990s), R (1990s), and computational scripting environments such as Matlab multiplied the paths along which machine learners move through operational formations. It would be difficult to comprehend the propagation of machine learners across domains of science, business, and government without paying attention to coding practices. Even if textbooks and research articles are not read, software packages and libraries for machine learning are used. Code has a mobility that extends the diagrammatic practices of machine learners into a variety of settings and places where the scientific reading apparatuses of equations, statistical plots, and citations of research articles would not be operative.

How should we think of the R code in *Elements of Statistical Learning* in its operational specificity? Its growth is perhaps just as important as its operation (Mackenzie 2014b). An open source programming language, according to surveys of business and scientific users, at the time of writing, R has replaced popular statistical software packages such as SPSS, SAS, and Stata as the statistical and data analysis tool of choice for many people in business, government, and sciences ranging from political science to genomics, from quantitative finance to climatology (RexerAnalytics 2015). Developed in New Zealand in the mid-1990s and like many open source software projects, emulating S, a commercialized predecessor developed at AT & T Bell Labs during the 1980s, R is now extremely widely used across life and physical sciences, as well as quantitative social sciences. John Chambers, the designer of S, was awarded the Association for Computing Machinery (ACM) "Software System Award" in 1998 for "the S system, which has forever altered how people analyze, visualize, and manipulate data" (ACM 2013). Many undergraduate and graduate students today earn R as a basic tool for statistics. Skills in R are often

---

where, presumably, the bulk of the readers work or aspire to work. Given a certain predisposition (i.e., geekiness), these books and other learning materials can make for enjoyable reading. Although they vary highly in quality, it is sometimes pleasing to see the economical way in which they solve common data problems. This genre of writing about programming specialises in posing a problem and solving it directly. In these settings, learning takes place largely through following or emulating what someone else has done with either a well-known dataset (Fisher's iris) or a toy dataset generated for demonstration purposes. The many frictions, blockages, and compromises that often affect data practice are largely occluded here in the interests of demonstrating the neat application of specific techniques. Yet are not those frictions, neat compromises, and strains around data and machine learning precisely what we need to track diagrammatically? To demonstrate both the costs and benefits of approaching R through such materials, rather than through ethnographic observation of people using R, it will be necessary to stage encounters here with data that have not been completely transformed or cleaned in the interests of neatly demonstrating the power of a technique. One way to do this is by writing about code while coding.

seen as an essential prerequisite for scientific researchers, especially in the life sciences. (In engineering, `Matlab` is widely used.) Research articles and textbooks in statistics commonly both use `R` to demonstrate methods and techniques and create `R` packages to distribute the techniques and sample data. Nearly all of these publication-related software packages, including quite a few from the authors of *Elements of Statistical Learning*, are sooner or later available from the "Comprehensive `R` Archive Network (CRAN)" (CRAN 2010). Estimates of its number of users range between 250,000 and 2 million. Increasingly, `R` is integrated into commercial services and products (e.g., SAS, a widely used business data analysis system now has an `R` interface; Norman Nie, one of the original developers of the SPSS package heavily used in social sciences, now leads a business, Revolution, devoted to commercializing `R`; `R` is heavily used at Google, at FaceBook, and by quantitative traders in hedge funds; in 2013 'R usage is sky-rocketing'; etc.). In general terms, `R` has currency as a form of polyglot expression and exhibits a fine-grained relationality with many different epistemic and operational situations associated with machine learning.

Two early proponents of `R` and `S` describe the motivation for the language:

> The goal of the S language . . . is "to turn ideas into software, quickly and faithfully" . . . it is the duty of the responsible data analysts to engage in this process . . . the exercise of drafting an algorithm to the level of precision that programming requires can in itself clarify ideas and promote rigorous intellectual scrutiny. . . . Turning ideas into software in this way need not be an unpleasant duty. (Venables and Ripley 2002, 2)

Bill Venables and Brian Ripley, statisticians working on developing `S`, the almost identical commercial predecessor to `R`, wrote in the early 1990s of the responsibility of data analysts to write, not just use, software. They "write software" here not in the sense of producing a product, but in the sense that today would more likely be called "code." Their sense of coding and programming as clarifying and concretizing ideas with precision—as abstractions—has thoroughly taken hold in contemporary data analysis where analysis is done through coding. If code, as they suggest, entails a threshold of idealization, it differs from mathematical formalization in that it changes the positions and relations of knowledge to include machines, devices, and infrastructures.

The view that code expresses ideas precisely does not capture the relational complexity of `R` code as it operates in a setting such as *Elements of Statistical Learning*. Code, along with mathematics, is a primary operational form for machine learners. But neither a view of code as an expressive operation by which "an individual formulates an idea" nor a view of it as "a rational activity that may operate in a system of inference" (Foucault 1972, 117) encompasses operational practice in machine learning. Similarly, the intersection of `R` with machine learning also lies somewhat at odds with Pedro Domingos'

Listing 2.3
Install packages in R

```
install.packages('ElemStatLearn', dependencies='Suggests', repos =
'http://cran.us.r-project.org')
```

characterization of machine learning as a shift away from people building to learners growing programs.

What in R (let alone other programming languages) overflows both the ideas of code as expression of ideas and code as automation? Alongside expression and automation, much R code traces a matrix of practice crossing network infrastructures, display screens, statistical techniques, software engineering architectures, as well as publication and documentation standards. For instance, the line of R code shown in the listing 2.3 when executed opens another way of reading *Elements of Statistical Learning* and sensing the heft of practical relations and infrastructural configurations running through it. Take the part of the line `dependencies = 'Suggests'`. When the line of code executes, the stipulation of `dependencies` triggers a wide-ranging operation that installs many R packages. If the installation works (and that assumes quite a lot of configuration work has already taken place; e.g., installing a recent version of the R platform), then *Elements of Statistical Learning* is now augmented by various pieces of code and by various datasets that in some ways echo or mimic the book but in other ways extend it operationally (see tables 2.5 and 2.4).[5] These code elements are often stunningly specialised. As Karl Marx wrote of the 500 different hammers made in Birmingham, "not only is each adapted to one particular process, but several varieties often serve exclusively for the different operations in one and the same process" (Marx 1986, 375). Something similar holds in R: thousands of software packages in

5 Most of the packages associated with the `ElemStatLearn` implement methods or techniques developed by Hastie, Tibshirani, or Friedman, but some are much more generic. `MASS`, for instance, is a highly cited R package. (Of the 8628 packages in the R CRAN system, 415 depend on the library `MASS`, itself an adjunct to the influential and highly cited *Modern Applied Statistics with S* (Venables and Ripley 2002), a textbook that presents many machine learning techniques using S, AT & T Bell Labs commercial precursor to the open sourced R.) For our purposes, this hardly accidental mixing of academic or research work with a programming languages and its associated infrastructures is fortuitous. It allows us to transit among different strata of the social fields of science, engineering, health, medicine, business media, and government more easily.

Table 2.4
R packages suggested by the ElemStatLearn package

	How often suggested
testthat	334
knitr	164
MASS	77
knitr, rmarkdown	74
testthat, knitr	52
testthat, knitr, rmarkdown	45
RUnit	37
knitr, testthat	30
parallel	26
R.rsp	26
lattice	25
knitr, rmarkdown, testthat	24
ggplot2	18
mvtnorm	16
survival	16
rgl	12
testthat, covr	9
xtable	9
testthat, roxygen2	8
akima	7

online repositories suggest that a highly specialized division of labor operates around data.

Because of this almost incoherent plurality, and its labile status as both a programming environment and a statistical analysis package, `R` is an evocative object, to use the psychoanalyst Christopher Bollas' term (Bollas 2008), an object through which many different ways of thinking circulate. Standing somewhere at the intersection of statistics and computing, modeling and programming, many different disciplines, techniques, domains, and actors intersect in `R`. It engages immediately, practically, and widely with words, numbers, images, symbols, signals, sensors, forms, instruments, and, above all, abstract forms such as mathematical functions like probability distributions and many different architectural forms (vectors, matrices, arrays, etc.) as it employs data. If, as Bollas suggests, "our encounter, engagement with, and sometimes our employment of, actual things is a *way* of thinking" (Bollas 2008, 92), then it plausible that `R` not only gathers a plurality of data practices—working with measurements, numbers, text, images, models, and equations, with techniques for sampling and sorting, with probability distributions and random numbers—but that it expresses the kernel of a machine learner mode of thought.

Table 2.5
R packages depended on by the "ElemStatLearn" package

	Package	How often depended on
methods	methods	74
stats	stats	33
survival	survival	31
MASS	MASS	26
mvtnorm	mvtnorm	21
Matrix	Matrix	13
ggplot2	ggplot2	12
lattice	lattice	9
rJava	rJava	9
grid	grid	7
igraph	igraph	7
tcltk	tcltk	7
XML	XML	7
rgl	rgl	6
graphics	graphics	5
methods, stats	methods, stats	5
nlme	nlme	5
utils	utils	5
vegan	vegan	5
ape	ape	4

The obdurate mathematical glint of machine learning

If scientific research literature and operational R code constitute the elements of an operational formation presented in *Elements of Statistical Learning*, then what of the mathematics? Although references from many different places flow in and out of *Elements of Statistical Learning*, they are nearly all articulated in mathematical form. Machine learning as a grouping of statements relies heavily on mathematics. Given that mathematics is diverse and multi-stranded, what kind of mathematics matters here? Although later chapters will explore some of the main mathematical practices (linear algebra, statistical inference, etc.), many of the machine learners in *Elements of Statistical Learning* coalesce around a single exemplary technique: linear regression models or fitting a line to points. The linear regression model is pivotal, not just in *Elements of Statistical Learning* but in much of the scientific and engineering literature. The linear regression model pushes up some of the citational peaks in figure 2.3. Even though it is an old technique dating back to Francis Galton in the 1890s (see (Stigler 1986, chapter 8)), it remains perhaps the central working element of machine learning in its latest transformations.

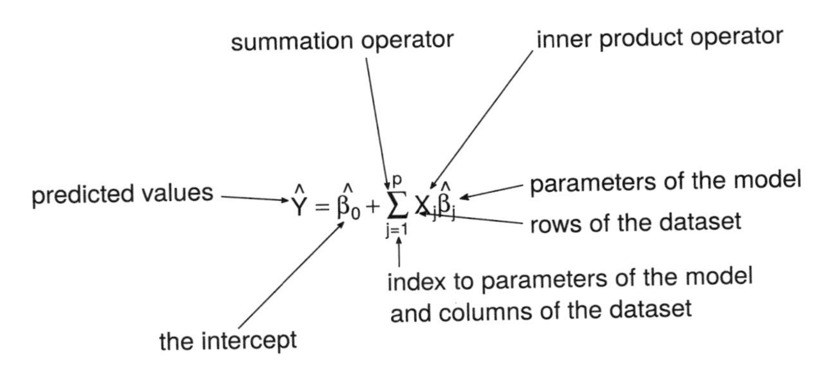

Figure 2.4
The linear regression model.

Elements of Statistical Learning acknowledges the statistical legacy and inheritance in machine learning:

The linear model has been a mainstay of statistics for the past 30 years and remains one of our most important tools. Given a vector of inputs $X^T = (X_1, X_2, \ldots, X_p)$, we predict the output Y via the model

$$\hat{Y} = \hat{\beta}_0 + \sum_{(j=1)}^{p} (X_j \hat{\beta}_j)$$

The term $\hat{\beta}_0$ is the intercept, also known as the *bias* in machine learning (Hastie, Tibshirani, and Friedman 2009, 11).

In the course of the book, this mathematical expression appears in many variations, iterations, expansions, and modifications ("ridge regression," "least angle regression," "project pursuit," etc.). However, this introduction of the "mainstay of statistics," the linear model, already introduces a diagrammatic element—the mathematical equation—that is perhaps the most prominent feature in the text.

Any reading of the book has to work out a way to traverse the forms show in equation 2.4. In its relatively compressed typographic weave, expressions such as equation 2.4 operationalize movements through data or "a vector of inputs X^T." These expressions, which are not comfortable reading for nontechnical readers, merit careful consideration if we want to move "at the same level of abstraction as the algorithm" (Pasquinelli 2015). They can be found in hundreds in (Hastie, Tibshirani, and Friedman 2009), but also in many other places. Their presence distinguishes machine learning from parts of computer science where mathematical equations are less common. Mathematical formalizations also allow the book to innervate scientific publications and datasets in fields of statistics, artificial intelligence, and computer science. Along with the citations, the graphical plots, and the code relationality, these equations are an integral connective tissue in machine learners.

In dealing with the obscuring dazzle of mathematical formalisation, I find it useful here to follow Charles Sanders Peirce's approach to mathematics. "Mathematical reasoning," he writes, "is diagrammatic" (Peirce 1998, 206). Peirce sees mathematics, whether it takes an algebraic or geometrical form, whether it appears in symbols, letters, lines, or curves, as diagrams. For Peirce, a diagram is a kind of "icon." The icon is a sign that resembles the object it refers to: it has a relation of likeness. What likeness appears in equation 2.4? "Many diagrams," Peirce writes, "resemble their objects not all in their looks; it is only in respect to the relations of their parts that their likeness consists" (Peirce 1998, 13). As we have seen in the perceptron code, machine learners can be expressed in statements in a programming language such as `Python` or `R`. In code, however, the relations between parts cannot be observed in the same way as they can in the algebraic form. The "very idea of the art," as Peirce puts it (Peirce 1992, 228), of algebraic expressions is that the formulae can be manipulated, or that component elements can be moved without much effort through substitutions, transformations, and variations. The graphic form of the expression includes the various classical Greek operator symbols such as $\sum$ or $\prod$, as well as the letters x, y, z and the indices (indexical signs) such as j that appear in subscript or superscript, as well as the spatial arrangement of all these elements in lines and sometimes arrays. A variety of relations run between these different symbols and spatial arrangements. For instance, in algebraic diagrams of machine learners, the relation between the left-hand side of the "=" and the right-hand side is important. By convention, the left-hand side of the expression is the value that is predicted or calculated (the "response" variable) and the right-hand side are the input variables or "features" that contribute data to the model or algorithm. This spatial arrangement fundamentally affects the design of algorithms. In the case of equations such as 2.4, the '^' over $\hat{Y}$ symbolizes a predicted value rather than a value that can be known completely through deduction, derivation, or calculation. This distinction between predicted and actual values organizes a panoply of different practices and imperatives (e.g., to investigate the disparities between the predicted and actual values—machine learning practitioners spend a lot of time on that problem).

The broad point is that the whole formulae is a diagram, or an icon that "*exhibits*, by means of the algebraical signs (which are not themselves icons), the relations of the quantities concerned" (Peirce 1998, 13). Because diagrams compress so many details, they allow one to focus on a selected range of relations between parts. The compression afforded by diagrammatic mathematical forms is extremely important in the operational formation of machine learning. Diagrams can diagram other diagrams. Operations can become the subject of operations. The nesting of diagrams is generative in that it allows what Peirce calls "transformations" (Peirce 1998, 212) or the construction of "a new general predicate" (Peirce 1992, 303).[6]

Although I seek to relate to the machine learner mathematics as diagrams and will present a selection of equations (nowhere near as many as found in *Elements of Statistical Learning*) in the following pages, I am not assuming their operation is transparent or fully legible. Just as much as the analysis of a photograph, a literary work or an ethnographic observation, their diagrammatic composition calls for repeated consideration. Peirce advises not to begin with examples that are too simple: "in simple cases, the essential features are often so nearly obliterated that they can only be discerned when one knows what to look for" (Peirce 1998, 206). He also suggests "it is of great importance to return again and again to certain features" (Peirce 1998, 206). Looking at these diagrammatic expressions repeatedly might allows us to map something of how transformations, generalizations, or intensification flow across disciplinary boundaries, across social stratifications, and, sometimes, generate potentially different ways of thinking about collectives, inclusion, and belonging.

CS229, 2007: Returning again and again to certain features

If we were to follow Peirce's injunction to "return again and again to certain features," how would we do that? *Elements of Statistical Learning* is a difficult book to read in isolation (although it does pay re-reading). Even after several years, the diagrammatic

6 Félix Guattari makes direct use of Peirce's account of diagrams as icons of relation in his account of "abstract machines" (Guattari 1984). He writes that "diagrammaticism brings into play more or less deterritorialized transsemiotic forces, systems of signs, of code, of catalysts and so on, that make it possible in various specific ways to cut across stratifications of every kind" (Guattari 1984, 145). Here the "transsemiotic forces" include mathematical formulae and operations (such as the banking system of Renaissance Venice, Pisa, and Genoa). They are transsemiotic because they are not tethered by the signifying processes that code experience or speaking positions according to given stratifications such as class, gender, nation, and so forth. Although Guattari (and Deleuze in turn in their co-written works (Guattari and Deleuze 1988)) is strongly critical of the way which signification territorializes (we might think of cats patrolling, marking, and displaying to maintain their territories), he is much more affirmative of diagrammatic processes. He calls them "a-signifying" to highlight their difference from the signifying processes that order social strata. He suggests that diagrams become the foundation for "abstract machines" and the "simulation of physical machinic processes." Writing in the 1960s, Guattari powerfully anticipates the abstract machines and their associated diagrams that have taken shape and physical form in the succeeding decades.

density of its "elements" or statements (equations, citations, tables, datasets, plots) leaves me with a refractory feeling of "not quite understanding." This feeling is inevitable because the book condenses finished work from several disciplines, and partly because it seeks to organize a great diversity of materials *epistemically*. Indeed, the book might seen as evidence that machine learning has crossed an epistemic threshold formulated in a statistical apparatus. (The statistical aspects of machine learning are the main topic of chapter 5.)

Does the pedagogical experience of a computer science course offer an easier route? Andrew Ng's course "Machine Learning" CS229 at Stanford (http://cs229.stanford.edu/) might provide a supplementary path into machine learning (Ng 2008a).[7] The course description runs as follows:

> This course provides a broad introduction to machine learning and statistical pattern recognition. Topics include supervised learning, unsupervised learning, learning theory, reinforcement learning and adaptive control. Recent applications of machine learning, such as to robotic control, data mining, autonomous navigation, bioinformatics, speech recognition, and text and web data processing are also discussed (Ng 2008a).

CS229 is in many ways a typical computer science pedagogical exposition of machine learning. Machine learning expositions usually begin with simple datasets and the simplest possible statistical models and machine learners (linear regression), and then, with a greater or lesser degree of attention to issues of implementation, move through a succession of increasingly sophisticated and specialized techniques. This pattern is found in many of the how-to books, in the online courses, and in the academic textbooks, including (Hastie, Tibshirani, and Friedman 2009).

Ng's CS229 lectures differ from most other pedagogical materials in that we see someone writing and deriving line after line of equations using chalk on a blackboard. Occasionally, questions come from students in the audience (not shown on the YouTube videos), but mostly Ng's transcription of equations and other diagrams from the paper notes he holds to blackboard continues uninterrupted.[8] Ng's YouTube anachronistic but popular lectures have a certain diagrammatic density that comes from the many hours of writing he chalks up during the course of deriving equations. In

7 A heavily shortened version of this course has been delivered under the title "Machine Learning" on Coursera.org, a MOOC (Massive Open Online Course) platform.

8 After sitting through 20 hours of Ng's online lectures, and attempting some of the review questions and programming exercises, including implementing well-known algorithms using `R`, one comes to know datasets such as the San Francisco house price dataset and Fisher's `iris` (Fisher 1936) quite well. Like the textbook problems that the historian of science Thomas Kuhn long ago described as one of the anchor points in scientific cultures (Kuhn 1996), these iconic datasets

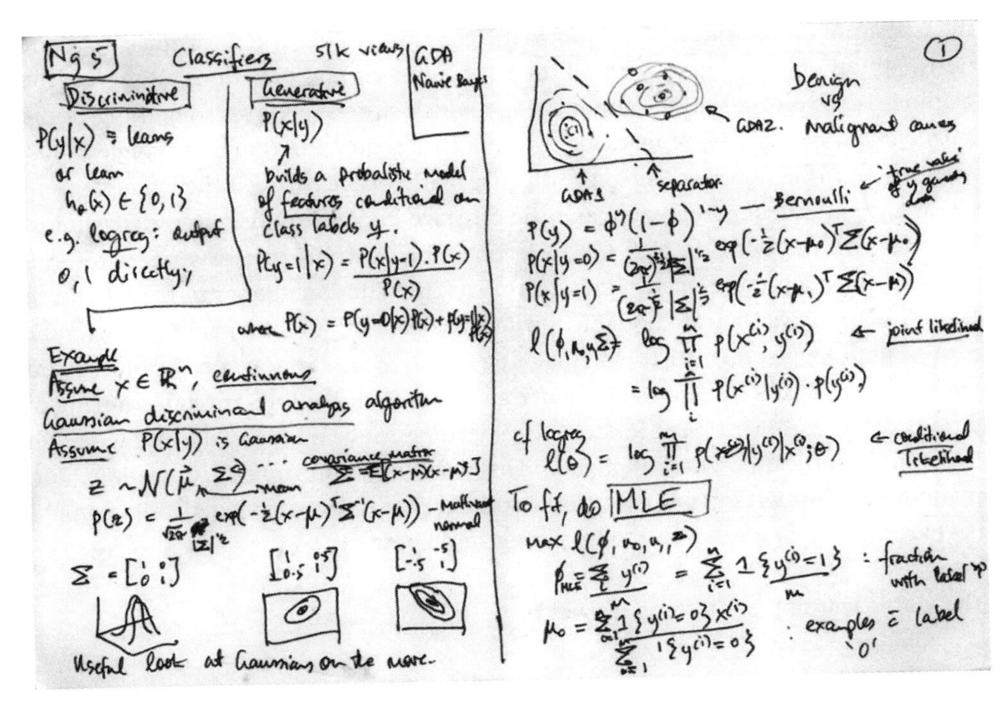

Figure 2.5
Class notes lecture 5, Stanford CS229, 2007.

a time when PowerPoint presentations or some other electronic textuality would very much have been the norm (2007), what was a computer science professor, teaching a fairly advanced postgraduate course, doing writing on a chalkboard by hand?

Figure 2.5 shows a brief portion of around 100 pages of notes I made on this course. The act of writing down these equations and copying the many hand-drawn graphs Ng produced was a deliberative diagrammatic experiment in "returning again and again" to what is perhaps overly hardened in *Elements of Statistical Learning*. Like the 50,000 or so other people who had watched this video, I partly complied with Ng's injunction to "copy it, write it out, cover it, and see if you can reproduce it" (Ng 2008d). Although it occasions much writing and drawing, and many struggles to keep up with

provide an entry point to the "disciplinary matrix" of machine learning. Through them, one gains some sense of how predictive models are constructed, and what kinds of algorithmic architectures and forms of data are preferred in machine learning.

the transformations and substitutions that Ng narrates as he writes, the quasi-mimetic transcription of derivations, with all their substitutions and variations, alongside the graphic sketches of intuitions about the machine learners, accesses something of the *diagrammatic composition* of machine learning that is quite hard to disentangle in *Elements of Statistical Learning*. There the diagrammatic weave between the expressions of linear algebra, calculus, statistics, and the off-stage implementation in code is almost too tight to work with. In Ng's CS229 lectures, by contrast, the weaving, while still complex, is much more open. The lectures lack the citational tapestry of *Elements of Statistical Learning*. They are not able to wield the datasets and the panoply of graphic forms found there, and virtually no machine learning code appears on the blackboard (although the CS229 student assignments, also to be found online, are code implementations of the algorithms and models). Ng's lectures move more slowly, and we begin to see some of the different groupings and associations comprising the operational formation. More important, perhaps, this absorbing process of writing derivations might begin to transform ways of thinking, saying, and knowing.

The visible learning of machine learning

For a book with "learning" in its title, *Elements of Statistical Learning* has visibly little to say about how to learn machine learning. Learning is briefly discussed on the first page of *Elements of Statistical Learning*, but the book hardly ever returns to the topic or even that term explicitly. We learn on page 2 that a "learner" in machine learning is a model that predicts outcomes. (As I discuss in chapter 4, learning is comprehensively understood in machine learning as finding a mathematical *function* that could have generated the data and optimizing the search for that function as much as possible.) The notion of learning in machine learning derives from the field of artificial intelligence. The broad project of artificial intelligence, at least as envisaged in its 1960s-1970s heyday as a form of symbolic reasoning, is today largely regarded as a dead end.

How then did learning get into the title of *Elements of Statistical Learning*? Does learning anthropomorphize statistical modeling or computer programming? The so-called "learning problem" and the theory of learning machines developed by researchers in the 1960–1970s was largely based on work already done in the 1950s on cybernetic devices such as the perceptron, the prototypical neural network model developed by the psychologist Frank Rosenblatt in the 1950s (Rosenblatt 1958). Drawing on the McCulloch-Pitts model of the neurone and mathematical techniques of optimization (Bellman 1961), Rosenblatt implemented the perceptron, which today would be called a single-layer neural network (Hastie, Tibshirani, and Friedman 2009, 393), as an

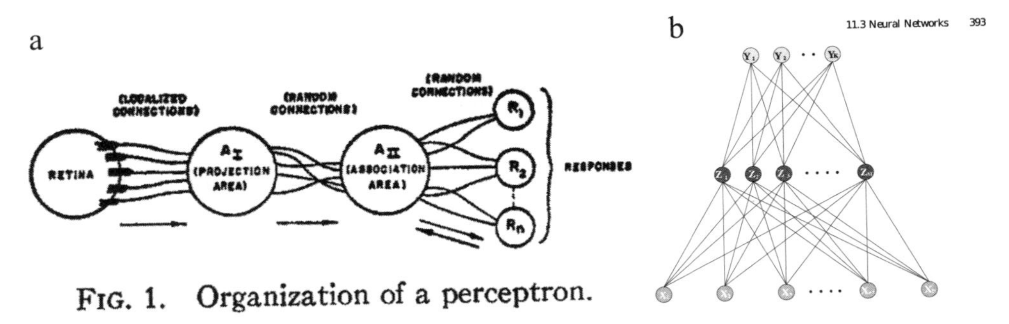

Figure 2.6
1958 perceptron and 2001 neural net compared. (a) The neurological perceptron (Rosenblatt, 1958, 389). (b) Neural network from Hastie (2009); single-layer feedforward artificial neural network.

electromechanical device at the Cornell University Aeronautical Laboratory in 1957. For present purposes, it is interesting to see what diagrams Rosenblatt used and how they differ from contemporary diagrams.

If we compare the diagram of Rosenblatt's perceptron shown in figure 2.6a to the typical contemporary diagram of a neural network shown in figure 2.6b, the differences are not that great in many ways. The diagram of a neural network found in Rosenblatt's paper (Rosenblatt 1958) has no mathematical symbols on it, but the one from (Hastie, Tibshirani, and Friedman 2009) does. Rosenblatt retains neurocognitive-anatomical reference points (retina, association area, projection area), whereas Hastie et al. replace them with the symbols that we have already seen in play in the expression for the linear model shown in figure 2.4. What has happened between the two diagrams? As Vladimir Vapnik, a leading machine learning theorist, observes: "the perceptron was constructed to solve pattern recognition problems; in the simplest case this is the problem of constructing a rule for separating data of two different categories using given examples" (Vapnik 1999, 2). Although computer scientists in artificial intelligence of the time, such as Marvin Minsky and Seymour Papert, were skeptical about the capacity of the perceptron model to distinguish or "learn" different patterns (Minsky and Papert 1969), later work showed that perceptrons could "learn universally."[9]

9 In describing the entwined elements of machine learning techniques and citing various machine learning textbooks, I'm not attempting to provide any detailed history of their development. My archaeology of these developments does not explore the archives of institutions, laboratories, or companies where these techniques took shape. It derives much more from either following citations back out of the highly distilled textbooks into the teeming collective labor

For present purposes, the key point is not that neural networks have turned out several decades later to be extremely powerful algorithms in learning to distinguish patterns and that intense research in neural networks has led to their ongoing development and increasing sophistication in many "real-world" applications (see e.g., for their use in sciences (Hinton and Salakhutdinov 2006) or in commercial applications such as drug prediction (Dahl 2013) and above all in the current surge of interest in "deep learning" by social media platform and search engines such as Facebook and Google). For our purposes, the important point is that the notion of the learning machine sets in motion an ongoing diagonalization or sideways slippage that transforms the basic diagram of the linear model through substitutions of increasingly convoluted or nested operations. The whole claim that machines "learn" rests on these diagrammatic substitutions that recurrently and sometimes recursively transform the icon of relations, sometimes in the graphic forms shown above but more often in the algebraic patterns.

The changes in graphics suggest transformations in the operational formation. I am suggesting, then, that we should follow the transformations associated with machine

---

of research on machine learning as published in hundreds of thousands of articles in science and engineering journals or looking at, experimenting with, and implementing techniques in code. For instance, (Olazaran 1996) offers a history of the perceptron controversy from a science studies perspective. During the 1980s, artificial intelligence and associated approaches (expert systems, automated decision support, neural networks, etc.) were a matter of some debate in the sociology and anthropology of science. The work of Harry Collins would be one example of this (Collins 1990), and Paul Edwards on artificial intelligence and Cold War (Edwards 1996), Nathan Ensmenger on chess (Ensmenger 2012), and Lucy Suchman on plans and expert systems (Suchman 1987) would be others. Philosophers such as Hubert Dreyfus (*What Computers Can't Do* Dreyfus 1972, 1992) had already extensively criticized AI. In the 1990s, the appearance of new forms of simulation, computational fields such as a-life, and new forms of robotics such as Rodney Brook's much more insect-like robots at MIT attracted the interest of social scientists (Helmreich 2000), among many others. Sometimes this interest was critical of claims to expertise (Collins) and at other times interested in how to make sense of the claims without foreclosing their potential novelty (Helmreich). By and large, I don't attend to controversies in machine learning as a scientific field, and I don't directly contest the epistemic authority of the proponents of the techniques. I share with Lucy Suchman an interest in how the "effect of machine-as-agent is generated" and in how "translations . . . render former objects as emergent subjects" (Suchman 2006, 2). I diverge around the site of empirical attention. I'm persevering with the diagrams in each of the following chapters to track the movement of tendencies that are not so visible in terms of either the controversies or assumptions of agency embodied in many AI systems of the 1980s. The agency of machine learning, in short, might not reside so much in any putative predictive or classificatory power if manifests, but rather its capacity to mutate and migrate across contexts.

learning diagrammatically, provided we maintain a rich understanding of diagram and remain open to multiple vectors of abstraction. Following Peirce, we might begin to see machine learning as a diagrammatic practice in which different semiotic forms—lines, numbers, symbols, operators, patches of color, words, images, marks such as dots, crosses, ticks, and arrowheads—are constantly connected, substituted, embedded, or created from existing diagrams. The diagrams we have already seen from *Elements of Statistical Learning*—algebraic formulae and network topology—don't exhaust the variations at all. Just a brief glance through this book or almost any other in the field shows not only many formulae but tables, matrices, arrays, line graphs, contour plots, scatter plots, dendrograms, and trees, as well as algorithms expressed as pseudocode. The connections among these diagrams are not always tight or close. Learning to machine learn (whether you are a human learner or a learner in the sense of a machine) means stepping among diagrams. The steps are relatively small and sometimes almost infinitesimal as signs slide between different diagrammatic articulations. To understand what machines can learn, we need to look at how they have been drawn, designed, or formalized. But what in this work of designing and formalising predictive models is like farming? Some divergent trajectories open up here. On the one hand, the diagrams become machines when they are implemented. On the other hand, the machines generate new diagrams when they function. We need to countenance both forms of movement to understand any of the preceding diagrams—the code, the algebraic expressions or diagrams of models, such as the perceptron or its descendants, the neural network. This means going downstream from the textbooks into actual implementations and places where people, algorithms, and machines mingle more than they do in the relatively friction-free neatness of the textbooks. Rather than history or controversies in the field, I focus on the migratory patterns of methods and the many configurational shifts associated with their implementations as the same things appears in different places.

The diagram of an operational formation

Does machine learning radically change programming practice? Programmability does change but only through gyrations among epistemic, infrastructural, and discursive heterogeneities practiced and practiced in code. I have been suggesting that we can get a sense of the heterogeneities and regularities of machine learning by treating *Elements of Statistical Learning* as a diagram that links and indexes the operations of machine learners in publication, computation, and code. Mapped through research publications, pedagogical materials, or code libraries in `R`, these operations form a primary field of expressions issuing from many parts. These parts include the accumulation or positivity of scientific literature with all the richness of its referentiality. They include mathematical derivation and formalization as a compressed diagrammatic movement. The parts cascade through code in the fluid weaving of infrastructures, conventions, standards, techniques, devices, and collective relations. Statements (especially linear algebra, calculus, statistics), problems, and techniques from multiple scientific disciplines (especially computer science, but also biology, medicine, and others), devices such as computing platforms, data formats, and code repositories populate the operational formation. The operational power of machine learning does not stem from a single layer of abstraction. Precisely the opposite: the diagrammatic forms of movement we have begun to discern in the polymorphic *Elements of Statistical Learning* point in several directions at once. Like the perceptron calculating weights that allow it to express the logic of the NAND function, we might first of all move to the table the ordering and aligning of numbers on which nearly all machine learning depends.

3 Vectorization and Its Consequences

The table has the function of treating multiplicity itself, distributing it and deriving from it as many effects as possible (Foucault 1997, 149).

We call *any* set that satisfies these properties (or axioms) a *vector space*, and the objects in the set are called *vectors* (Larson 1996, 166).

All things are vectors (Whitehead 1960, 309).

Machine learning locates data practice in an expanding epistemic space. The expansion derives, I will suggest, from a specific operational diagram that maps data into a vector space. It *vectorizes* data according to axes, coordinates, and scales. Machine learners, in turn, inhabit a vectorized space, and their operations vectorize data.

Often data are represented as a homogenous set of numbers or a continuous flowing stream. We need, however, to archaeologically examine some of the transformations that allow different shapes and densities of data, whether in the form of numbers, words, or images, to become machine learnable. Data in their local complexes space out in many different density shapes, depending on how the changes, signals, propensities, and norms have been generated or configured.[1] Whatever the starting point—a measuring instrument, people clicking, and typing on websites, a device like a camera, a random number generator, and so on on—machine learners only ever encounter data in specific *vectorized* shapes (vectors, matrices, arrays, etc.) mapped to a geometrically coordinate volume. The mapping and forming, when mentioned at all, is sometimes referred to as "data cleaning," but that term covers over important but largely

1 I loosely borrow the term "density" from statistics, where *probability density functions* are often used to describe the hardly ever uniform distribution of probabilities of different values of a variable. Sensing density as a form of variation matters greatly in machine learning, where algorithms seek purchase on unevenly distributed data, and in any broader diagram of data. Probability densities are discussed in much more detail in chapters 5. The collection *"Raw Data" is an Oxymoron* (Gitelman 2013) evinces some of these different densities and distributions of data.

taken for granted and constitutive transformations. The archaeology of data shapes presented in this chapter explores a range of transformations focused around the table or row-column grid.

The reshaping and reflowing of data densities into vectors deeply affects machine learning. This forming and reforming of data is evidence of implicated relations. The philosopher A.N. Whitehead called "strain":

> a feeling in which the forms exemplified in the datum concern geometrical, straight, and flat loci will be called a "strain." In a strain, qualitative elements, other than the geometrical forms, express themselves as qualities implicated in those forms (Whitehead 1960, 310).

For many machine learners, the exemplified forms are straight or flat loci (as we see in chapter 4). Yet different practices also elicit relations that strain the linear shaping of data, and these divergent relations sometimes combine in generative and provocative ways.

Vector space and geometry

Statistical modeling, data-mining, pattern recognition, recommendation systems, network modeling, and machine learning rely on the operation called "fitting a model." Fitting, or finding a good position, has some elements that resemble the phenomenologist Edmund Husserl's account of the origin of geometry. Husserl writes:

> First to be singled out from the thing-shapes are surfaces—more or less "smooth," more or less perfect surfaces; edges, more or less rough or fairly "even"; in other words, more or less pure lines, angles, more or less perfect points; then again, among the lines, for examples, straight lines are especially preferred, and among surfaces, the even surfaces.... Thus the production of even surfaces and their perfection (polishing) always plays its role in praxis (Derrida 1989, 178).

Husserl is attempting to describe something of the way in which forms such as planes, lines, circles, triangles, squares, and points became objects of geometrical practice. A similar polishing and smoothing of surfaces is certainly taking place today in the thing-shapes we call data. The basic machine learning work of "fitting a model" (or many models) to data is often literally implemented, as we will see, by constraining data within a coordinate, discretized space, which I term the *vector space*.

Critical thought from phenomenology to social theory has a long-standing nervousness about the power of geometry and its gradual movement away from shapes and things toward mathematical operations. The philosopher Hannah Arendt, for instance, observes:

> decisive is the entirely un-Platonic subjection of geometry to algebraic treatment, which discloses the modern idea of reducing terrestrial sense data and movements to mathematical symbols (Arendt 1998, 265).

The crux of the problem rests on the "treatment" or operations that "reduce terrestrial sensibilities and movements" to symbols. One challenge for contemporary thought is how to orient itself to such operations, particularly in their data-intensive forms, without assuming that the familiar story of scientific and technical reduction of sense and movement to the lines and planes of modern geometry is simply reinforced in contemporary data practice. If an archaeology of data vectorization does anything, it needs to offer an alternative to that reduction.

Vectorizing operations also, as we will see, reach down into the practices of programming, infrastructure, and hardware production in ways that differ somewhat from increases in computational power or speed. Familiar narratives of Moore's Law increases in speed or efficiency of computation do not account for transformations of data in `R` and other computing environments (e.g., the popular `Map-Reduce` architecture invented at Google Corporation to speed up its search engine services (Mackenzie 2011)). Vectorization transforms data along more diagrammatic lines.[2]

Mixing places

Data appears in *Elements of Statistical Learning* in multiple forms. Maps of New Zealand fishing patterns lie next to plots of factors in South African heart disease. The 21 datasets shown in table 3.1 typify the variety found in Hastie, Tibshirani, and Friedman (2009). They span scientific, clinical, commercial, and media fields. Note that they include many patches and pathways of everyday life—speaking, seeing, writing, and reading—as well as specific scientific objects of knowledge, such as galaxies, cancer, climate, and national economies. This mixture is not unusual for machine learners. To give another example, statistician Leo Breiman's 2001 article on random forests, perhaps the most cited journal paper in the machine learning literature, displays a similarly diagonal line across human and nonhuman worlds:

> Glass, Breast cancer, Diabetes, Sonar, Vowel, Ionosphere, Vehicle, Soybean, German credit, Image, Ecoli, Votes, Liver, Letters, Sat-images, Zip-code, Waveform, Twonorm, Threenorm, Ringnorm (Breiman 2001b, 12).

2 Later chapters will discuss various ways in which the vectoral dimensionality of data or its rendering as *vector space* scales up and down in machine learning. In terms of multiplying matrices, dimensionality both constrains and enables many aspects of the prediction. Perhaps on the grounds of data dimensionality alone, we should attend to dimensionality practices in `R`. A fuller discussion of dimensionality of data is the topic of chapter 6. There I discuss how machine learning redimensions data in various ways, sometimes reducing dimensions and at other times multiplying dimensions.

Table 3.1
Datasets in *Elements of Statistical Learning*

	Item	Title
1	SAheart	South African Hearth Disease Data
2	bone	Bone Mineral Density Data
3	countries	Country Dissimilarities
4	galaxy	Galaxy Data
5	marketing	Market Basket Analysis
6	mixture.example	Mixture Example
7	nci	NCI microarray data (chap 14)
8	orange10.test	Simulated Orange Data
9	orange10.train	Simulated Orange Data
10	orange4.test	Simulated Orange Data
11	orange4.train	Simulated Orange Data
12	ozone	Ozone Data
13	phoneme	Data From a Acoustic-Phonetic Continuous Speech Corpus
14	prostate	Prostate Cancer Data
15	spam	Email Spam Data
16	vowel.test	Vowel Recognition (Deterding data)
17	vowel.train	Vowel Recognition (Deterding data)
18	waveform.test	Simulated Waveform Data
19	waveform.train	Simulated Waveform Data
20	zip.test	Handwritten Digit Recognition Data
21	zip.train	Handwritten Digit Recognition Data

In some ways, the improbable conjunction of spam email and cancer detection in machine learning continues what statistics as a field has always done: rove across scattered fields ranging from astronomy to statecraft, from zoology to epidemiology, gleaning data as it goes (see Stigler 1986; Hacking 1990, for samples of itineraries).

In *Elements of Statistical Learning* and the field of machine learning more generally, something aligns, connects, and coordinates these datasets. The coordination might be rhetorical: the colligation of datasets—vowels, ozone, bone density, marketing, prostate cancer, and spam—in all their diversity suggests the mobility of machine learners. The combination of datasets deriving from network media, medicine, business administration, and cutting-edge life science (c. 2000) suggests a tremendous, indeed almost spectacular, miscibility, one that in principle could surprise us because otherwise little mixing occurs between the settings and knowledge domains these datasets come from. How is this mixing, conformation, and homogenization being done? The repetition of datasets, juxtaposition of discontinuous domains, and forms of movement construct and order continuities in the service of various forms of predictive and inferential knowledge. The miscible juxtapositions we encounter in machine learning enter into, it seems, a *regularity* or a common space, a space that displays strong tendencies to expand, accumulate, and archive relations. Rather than peripatetic learning, the accumulation of diverse datasets attests to a prior ordering of data to afford its traversal.

The practices of naming and ordering, the sorting of different data types, and the addressing and expansion of data exemplified in these diverse datasets can tell us a lot about how machine learners address data. The shapes, compositions, and loci formed from the datasets enable the functioning of machine learners as they operate to generate statements, classifications, and decisions. Machine learning can only evolve practical, diagrammatic abstractions amid the contours and features of diverse data. As we will see, machine learners transpose data in an increasingly extensive, heavily coordinated space, the vector space.

Truth is no longer in the table?

"This book is about learning from data," write Hastie, Tibshirani, and Friedman on the first page of *Elements of Statistical Learning*, as they rapidly begin to iterate through some datasets. On the second page of the book, a table of spam email word frequencies appears (and the problem of spam classification is canonical in the machine learning literature; we return to it in chapter 5). They come from the dataset `spam` (Cranor and LaMacchia 1998). On the third page, a complicated data graphic appears (figure 1.1; Hastie, Tibshirani, and Friedman 2009, 3). It is a scatterplot matrix of the `prostate` dataset included in the `R` package `ElemStatLearn`, the companion `R` package for the book. In a third example, a set of scanned handwritten numbers appears. These scans are images of zipcode or postcode numbers written on postal envelopes taken from the dataset `zip` (LeCun and Cortes 2012), and they differ from both the `spam` table and `prostate` plots because they directly resemble something in the world, which, however, happens to be numbers and is, therefore, probably already recruited into data-making and data-circulating processes (we return to a dataset in chapter 8). The final example in the introduction, "Example 4: DNA Expression Microarrays," draws from biology, and particularly high-throughput genomic biology, a science that produces large amounts of data about biological processes by running many tests or constructing devices that generate many measurements, in this case, a DNA microarray.[3]

3 Some genomic data will be the focus of chapter 7. Machine learning during the 1990s and 2000s was in some ways boosted heavily by the advent of genomic biology with its large, enterprise-style knowledge endeavors such as the Human Genome Project.

The table or row-column addressable grid is common to all of these datasets. Yet, as we are about to see, machine learning in many ways deals with the collapse or liquidation of tabular datasets. "Things, in their fundamental truth," writes Foucault in *The Order of Things*, "have now escaped from the space of the table" (Foucault 1992 [1966], 239). Foucault writes in these pages about the fabled emergence of life, labor, and language as the anchoring vertexes of a new triangle of knowledge and power structuring the figure of the "human" in the 19th century.

Before the emergence of the characteristic sciences of the human—political economy, linguistics, and biology—knowledges such as natural history, the general grammars, and philosophies of wealth (such as Adam Smith's work) had ordered empirical materials of diverse provenance in tables or grids. Although the history of tables as data forms reaches a long way back (see Marchese 2013 for a broad historical overview that reaches back to Mesopotamia), Foucault argues that the Classical Age first developed the system of grids that permitted ranking, sorting, and ordering in tables. These grids replace the Renaissance tabulations based on "buried similitudes" and "invisible analogies" (Foucault 1992 [1966], 26). In pre-Classical tables, an image or a figure from myth might lie alongside a measurement or count of occurrences, and this proximity was ordered by systems of analogical association that spanned what we might today, in the wake of the 19th century, see as incongruous (e.g., associations between medicine and Biblical prophecy).

The Classical Age grid or table, by contrast, brought plural and diverse resemblances into exhaustive systematic visible enumeration. As Foucault puts it:

> The space of order, which served as a *common place* for representation and for things, for empirical visibility and for the essential rules, which united the regularities of nature and the resemblances of imagination in the grid of identities and differences, which displayed the empirical sequence of representations in a simultaneous table, and made it possible to scan step by step, in accordance with a logical sequence, the totality of nature's elements thus rendered contemporaneous with one another (Foucault 1972, 239).

The table as space of order did not stand in isolation. It served a localized epistemic function in conjunction with other knowledge, such as experiment and mathematical proof. Because algebra or experimental *mathesis* only applied to "simple natures" (planets in movement, dynamics of falling bodies, etc.), table-based knowledges such as taxonomy dealt with more complex natures. In such tables, systems of signs (e.g., the groupings established by the 18th-century taxonomist Carl Linnaeus) sought to reduce complex natures (plants, animals, etc.) to simpler forms as columns and rows in a table based on resemblances and similarities. Importantly, the table as space of order was a space of imagination, in that one could begin to see continuities and differences between things (organisms, words, nations) by carefully ordering and scanning the table. "Hedged in by calculus and genesis" Foucault suggests, "we have the area of the *table*" (Foucault 1992 [1966], 73). Note in passing that calculus and calculations approach or enter the table only in relation to "simple natures" whose identity and difference can be understood in the form of movements, rates, and change in position.[4] This limitation is important because it is precisely the complex natures in genesis that machine learning tries to engage using algebra, calculus, statistics, and computation.

Finally (at least for our purposes), in the 19th century, a different form of ordering cut across tables based on enumerated similarity and ordered resemblance. Foucault figures this change as shattering the table into shards of order:

> this space of order is from now on shattered: there will be things, with their own organic structures, their hidden veins, the spaces that articulates them, the time that produces them; and then representation, a purely temporal succession, in which those things address themselves (always partially) to a subjectivity (Foucault 1992 [1966], 239–240).

Life, labor, and language—Foucault's famous historically emergent triadic figure of the human—replace the enumerative, synoptic classificatory tables of the Classical Age. Tables still abound in newer temporal, genetic orderings (almost any episode from the history of 19th- and 20th-century statistics will confirm that; see Stigler 1986, 2002), but from now on tables are localized, relating to a place and changing in time and functioning as shards of representational order. A double interiority takes root. Things such as a language, a species, or an economic system have their own genesis, temporality, and historical existence. Simultaneously, our knowledges and indeed experience of them become finite, historical, and with their own dynamics and internal life. In this change, the table is no longer the foundation or distillation of knowledge. It is one representational configuration among many. Tables, we might say, become merely data, inscriptions dependent on hidden structures and their genesis.

This brief résumé of a thread of argument in *The Order of Things* might help us reassess how machine learning goes into the data. The tables of `spam`, `prostate`, `image`, and `microarray` data in contemporary machine learning do not operate in the way Linnaeus would have tabulated a table of living things based on similarities and resemblances, or a Renaissance medical textbook might assemble analogical resemblances between disease and astronomy. As I will argue, measures of similarity and resemblance

4 In *Surveiller et Punir* or *Discipline and Punish* (Foucault 1977), Foucault returned to the question of the table in a slightly different context: an account of the operation of power in disciplinary institutions and knowledges. In these knowledges, the table becomes generative of "as many effects as possible."

still operate strongly in machine learning as it moves through tables. In this respect, the Classical table and perhaps even the pre-Classical table returns with extended relevance. The repeated juxtaposition of tables of diverse provenance alongside each other, the incorporation of "complex natures," and the operational superimposition of different tables in contemporary data practice suggest that machine learners still seek synoptic alignments but along different lines than the row-column grid of the Renaissance, Classical, or even the later life-labor-language tables of the human sciences. Even if the tables in the opening pages of *Elements of Statistical Learning* concern work, life, language, and economy, and map readily onto the anchor points, the new "empiricities" (Foucault's term for the empirical problems), of labor, life, and language or biopower that took root at this time (Foucault 1991), the ways machine learners traverse them may not be recognizable in terms of these empiricities.

We are instead now confronted by kaleidoscopically transmuted tables whose expansion and open margins afford many formulations of similarity and difference. Already in the handwritten digits and the microarray data, scale, and dimensions thwart tabular display of the data. In settings such as social media platforms or genomics in which machine learning operates, tables change rapidly in scale and sometimes in organization. The multiplication and juxtaposition of tables suggests that we might be seeing the advent of a postorder space for regularities and resemblances, for simple and complex natures, encompassing knowledges ranging from humanities to traffic engineering.

The epistopic fault line in tables

The `ElemStatLearn` `R` package brings, as we have seen in the previous chapter, with it around 20 different datasets, including the four mentioned in the introduction to *Elements of Statistical Learning*. On the one hand, every data table indexes a localized complex of activities (clinical research, social media platform, financial transactions, etc.) with possible referential importance. On the other hand, for machine learners, the table is a space of potential similarities and differences both internal to the table (e.g., how much does row number 1000 differ from row 1,000,000?) and associated with other tables (e.g., how much does this table of clinical test relate to that table of microarray data?). These internal and external differences entwine with each other in ways that create a fault line, an unstable yet generative line of diagonal movement. This fault or fold line, I propose, diagrammatically bases data tables in the expansive and moving substrata of the vector space.

The table has been vectorized. We might say that the vectorization of the table is epistopic. The term "epistopic" comes from the work of the science studies scholar Mike Lynch, whose account of scientific practice is particularly focused on ordinariness. Akin to what Foucault in the *Archaeology of Knowledge* terms a threshold of epistemologization (Foucault 1972, 195), Lynch characterizes the "epistopic" as connecting localized practices ("topics") with "familiar themes from epistemology and general methodology" in the local achievement of coherence in knowledge (Lynch 1993, 280). In other words, as the term suggests, an epistopic connects general epistemic themes such as validity, precision, specificity, error, confidence, expectation, likelihood, uncertainty, or approximation with a place, a "local complex of activities" (281). This emphasis on epistemic location frames the problem of how the "local complex" of a specific dataset encounters a generalizing data practice such as machine learning.

Surface and depths: The problem of volume in data

The local complex of activities that arranges and aligns datasets in vector space works on tables from a different angle than that described by Foucault in *The Order of Things*. Algebra, linear algebra in particular, organizes and distributes differences in vector space. *Mathesis* in the form of algebraic operations of addition and multiplication of collections of tabular elements such as rows and columns, now redefined as vectors, remap tabular data as a vector space, as a "set" whose membership is only limited by the applicability of the constructing relations. These operations absorb and subtend differences in quality, type, kind, and quantity.

Of the three example datasets (`prostate`, `spam`, and `zip`), *Elements of Statistical Learning* returns most frequently to `prostate`. This dataset derives from the work of urologists working at Stanford (Stamey et al. 1989) and concerns various clinical measurements performed on men who were about to undergo radical prostatectomy. The measurements range across the volume and weight of the prostate, as well as levels of various prostate-related biomarkers such as prostate specific antigen (PSA). Several rows from the dataset are shown in table 3.2. The first pages of the book already exhaustively plotted all the variables in the dataset against each other using the table-related form of a scatter plot matrix (Hastie, Tibshirani, and Friedman 2009, 3) (shown in figure 3.1), and the authors return to the same data on almost a dozen occasions in the course of the book, subjecting it to repeated vectorization.

The contrast between table 3.2 and figure 3.1 already depends on a transformation intrinsic to vectorization. On the one hand, the table arrays all different data types in rows and columns. In the table, the relation between the different data types (e.g., the

Table 3.2
First rows of the "prostate" dataset

	lcavol	lweight	age	lbph	svi	lcp	gleason	pgg45	lpsa	train
1	−0.58	2.77	50	−1.39	0	−1.39	6	0	−0.43	TRUE
2	−0.99	3.32	58	−1.39	0	−1.39	6	0	−0.16	TRUE
3	−0.51	2.69	74	−1.39	0	−1.39	7	20	−0.16	TRUE

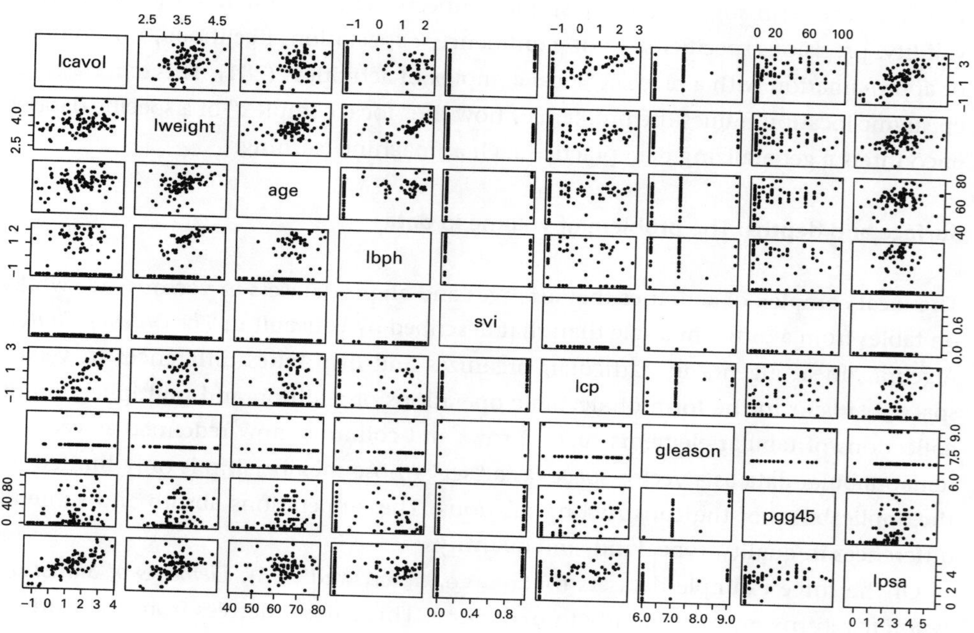

Figure 3.1
Scatter plot matrix of `prostate` data.

log of the weight of prostate - `lwp` and `age`) is quite hard to see. Moreover, different kinds of variables stand side by side. `svi`, short for seminal vesicle invasion, is a categorical variable. It takes the values true or false, shown here as `1` or `0`, but the other variables either measure or count things (e.g., years, sizes, or levels of antigens).

On the other hand, the scatter plot matrix also takes the form of a grid-like figure. Yet the cells of the grid are not occupied by numbers but by `x-y` plots of different pairs of variables in the `prostate` dataset. The matrix of figures shows of 72 plots is mirrored across the diagonal that runs from the top left to bottom right in figure 3.1. Taking this folding into account, we see 36 unique plots with different data in each one. Each subplot displays the relation between two variables in the dataset as a scatter plot. Some variables, such as `svi`, are not amenable to plotting in this way. More importantly, perhaps, certain combinations of variables appear as flat loci that can be read as signs of relations between different variables. The scatter plot matrix constructs a tabular space in which relational contrasts between pairs of variables start to appear. In the light of these contrasts (and I use light here in an almost literal sense to refer to the way in which the architecture of the figure creates a space in which light scatters in varying patterns), the `prostate` dataset begins to expose relations that might be worth knowing about. We have moved on from the bare table of the dataset to a transformed tabulation, from a textual-numerical grid to a geometrical-numerical grid. Everything remains on the surface of a grid here, but the grid permits differences in relationality to begin to appear.

All of this somewhat precedes the operational formation of machine learning. Similar tables and plots are part and parcel of statistical data exploration more generally (see Beniger and Robyn 1978 for a historical account of quantitative graphics in statistics). The scatterplot matrix does not exhaust differences or relationality in the `prostate` dataset but highlights a tendency to approach it from different angles (12 times in the *Elements of Statistical Learning*) to map the multiple relations or influences that remain opaque to even the most exhaustive matrices of plots. The scatterplot matrix shows pairs of variables in relation. If the crucial diagnostic factor in this case is an elevated level of the PSA, how do we know what combinations of other measurements might be associated with its elevation? What if multiple variables affect the level of PSA?[5] This question can be pursued by scanning the matrix of plots but not very stably because different data analysts might see different associations combining with each other there. Different statements or epistopics could be supported by the same figure.

The question of relation between multiple variables and the predicted levels of PSA suggests the existence of a hidden, occluded, or internal space that cannot be seen

5 Despite the intensive work that Hastie and coauthors conduct on the `prostate` data, all with a view to better predicting PSA levels using volumes and weights of prostates, Stamey and other urologists more than a decade concluded that PSA is not a good biomarker for prostate cancer. Stamey writes in 2004:

> What is urgently needed is a serum marker for prostate cancer that is truly proportional to the volume and grade of this ubiquitous cancer, and solid observations on who should and should not be treated which will surely require randomized trials once such a marker is available. Since there is no such marker for any other organ confined cancer, little is likely to change the current state of overdiagnosis (and over-treatment) of prostate cancer, a cancer we all get if we live long enough (Stamey et al. 2004, 1301).

in a data table and cannot be brought to light even in the more complex geometry of a plot. This volume contains the locus of multiple relations, a locus inhering in a higher dimensional space, in this case, the nine-dimensional space subtended by treating each of the nine variables or columns in the `prostate` dataset as occupying its own dimension. A different basis of order—the vector space—begins to take shape when dataset variables (usually columns in a table) become dimensions.[6]

Vector space expansion

To show how vector space expands, we might follow what happens to just one or two columns of the `prostate` data in the vector space as it is vectorized. In the `prostate` dataset, some variables are continuous quantitative values, some are categorical (they represent membership in a group or category), and some are ordinal variables (they represent a ranking or order). How can different data types be located in vector space? To put classifications or categories into vector space, they need to be translated into the same *basis* as the quantitative variables with their rather more obvious geometrical and linear coordinate values. How does one geometrically or indeed algebraically render a category or qualitative difference? The problem is solved via an expansion of the vector space through a form of binary coding that generates a new variable and hence a new dimension for each category:

> Qualitative variables are typically represented numerically by codes. The easiest case is when there are only two classes or categories, such as "success" or "failure," "survived" or "died." These are often represented by a single binary digit or bit as 0 or 1, or else by 1 and 1 . . . When there are more than two categories, several alternatives are available. The most useful and commonly used coding is via dummy variables. Here a K-level qualitative variable is represented by a vector of K binary variables or bits, only one of which is "on" at a time (Hastie, Tibshirani, and Friedman 2009, 12).

Again, the details are not so important here as the transformations that the vector space accommodates. A single qualitative or categorical variable expands into "a

6 Every distinct column in a table adds a new dimension to the vector space. Since the 1950s, problems of classification and prediction in high-dimensional spaces have been the object of mathematical interest. The mathematician Richard Bellman coined the term "the curse of dimensionality" to describe how partitioning becomes more unstable as the dimensions of the space increase (Bellman 1961). The problem is that while the volume of a space increases exponentially with dimensions, the number of data points (actual measurements or observations) usually does not usually increase at the same rate. In high dimensional spaces, the data becomes more thinly spread out. It is hard to partition sparsely populated spaces because they accommodate many different boundaries.

vector of K binary variables or bits." Qualitative data, once coded in this way, can be multiplied, added, and, in short, handled algebraically using the same aggregate operations applied to numerical or continuous variables. Not only has the vector space expanded here, its expansion smooths over important fault lines of difference that vertically divided the tabular data. Complex natures become simple natures. The different kinds of variables—qualitative and quantitative, discrete and continuous, nominal and ordinal—can be accommodated by adding dimensions to the vector space. As Whitehead says, "all things are vectors" (Whitehead 1960, 309).

Adding dimensions to vector space subsumes differences but makes seeing the geometrically regular loci—lines, planes, smooth surfaces—in data distributed in this space more challenging. The many transformations in `prostate` that ensue in *Elements of Statistical Learning* become the locus of machine learning. In a historically significant transfiguration of the table, these expansions—and we will see others, including *de novo* creations of constructed dimensions—subtend differences in a vector space comprising elements defined purely by coordinate position and vectoral (having direction and extent) movement. Once this hidden, expandable, and transformable (by rotation, displacement, or scaling) distribution of elements in space exists, strenuous efforts will be made to bring loci to light. Machine learners search for these loci or feel for data strains, to use Whitehead's term, along different lines. Sometimes a machine learner prehends vector space as filled with constantly varying proximities. It gathers and orders these proximities (e.g., as in the *k* nearest neighbors model) or in unsupervised methods such as k-means clustering (Hastie, Tibshirani, and Friedman 2009, 513). More commonly, machine learning draws lines or flat surfaces that constrain the volume.

The importance of lines and flat surfaces can hardly be under-estimated in machine learning. Finding lines of best fit underpins many of the machine learners that attract more attention (neural nets, support vector machines, random forests). Linear regression with its pursuit of the straight line or plane projects the basic alignments of vector space. It renders all differences as distances and directions of movement. Drawing lines or flat surfaces at various angles and directions is perhaps the main way in which the volume of data is traversed and a relation between input and output, between predictors and prediction, consolidated as a loci or data strain loci or data strain.[7] The

7 One sign of the centrality of the line in machine learning can be seen, for instance, from the contents page of the book (Hastie, Tibshirani, and Friedman 2009, xiii–xxii). After the introduction of the linear model in the first chapter and its initial exposition in chapter 2 ("overview of supervised learning"), it forms the central topics of chapter 3 ("linear methods for regression"), chapter 4 ("linear methods for classification"), chapter 5 ("basis expansions and regularization"), chapter 6 ("kernel smoothing methods"), much of chapter 7 ("model assessment and

line of best fit has a ready generalization to higher dimensions, and a line can be diagrammed in the equations of linear algebra, the field of mathematics that operates on lines in spaces of arbitrary dimensions. Linear algebra operations exist for finding intersections between lines and planes, manipulating collections of elements and aggregate forms such as matrices through mappings and transformations (rotations, displacements or translations, skewing, and scaling), and, above all, handling whole vector spaces as operational sets. It brings with it a set of formalizations—vector space, dimension, matrix, determinant, coordinate system, linear independence, eigenvectors and eigenvalues, inner-product space, and so on—that machine learners constantly and implicitly resort to invoke.[8]

Many of these operations quickly become difficult to geometrically figure.[9] Let us return to the equations for linear regression models (remembering that both C.S. Pierce

---

selection"), chapter 8 ("model inference and averaging"), major parts of chapter 9 ("additive models, trees, and related methods"), important parts of chapter 11 ("neural networks"—neural networks can be understood as a kind of regression model), the anchoring point of chapter 12 ("support vector machines and flexible discriminants"), and the main focus in the final chapter ("high-dimensional problems"). A similar topic distribution can be found in Andrew Ng's Cs229 lectures on machine learning. More than half of the 20 lectures concern linear models and their variants. See (Ng 2008b,c,d,e).

8 Along with statistics and probability, linear algebra is such an important part of machine learning that many books and courses recommend students complete a linear algebra course before they study machine learning. Cathy O'Neill and Rachel Schutt advise: "When you're developing your skill set as a data scientist, certain foundational pieces need to be in place first—statistics, linear algebra, and some programming" (Schutt and O'Neil 2013, 17).

9 We might also approach the epistopic fault line in machine learning topologically. More than a decade ago, the cultural theorist Brian Massumi wrote that "the space of experience is really, literally, physically a topological hyperspace of transformation" (Massumi 2002, 184). Much earlier, Gilles Deleuze had conceptualized Michel Foucault's philosophy as a topology, or "thought of the outside" (Deleuze 1988b), as a set of movements that sought to map the diagrams that generated a "kind of reality, a new model of truth" (Deleuze 1988b, 35). More recently, this topological thinking has been extended and developed by Celia Lury among others. In "The Becoming Topological of Culture," Lury, Luciana Parisi, and Tiziana Terranova suggest that "a new rationality is emerging: the moving ratio of a topological culture" (Lury, Parisi, and Terranova 2012, 4). In this new rationality, practices of ordering, modeling, networking, and mapping co-constitute culture, technology, and science (Lury, Parisi, and Terranova 2012, 5). At the core of this new rationality, however, lies a new ordering of continuity. The "ordering of continuity," Lury, Parisi, and Terranova propose, takes shape "in practices of sorting, naming, numbering, comparing, listing, and calculating" (4). The phrase "ordering of continuity" is interesting because we don't normally

and Andrew Ng advocate returning often to equations). The mainstay of statistics, the linear regression model, usually appears in a more or less algebraic form:

$$\hat{Y} = \hat{\beta}_0 + \sum_{j=1}^{p} X_j \hat{\beta}_j \tag{3.1}$$

$$\hat{Y} = X_T \hat{\beta} \tag{3.2}$$

Equations 3.1 and 3.2 express a plane (or hyperplane) in increasingly diagrammatic abstraction. The possibility of diagramming a high-dimensional space derives largely from linear algebra. Reading equation 3.1 from left to right, the expression $\hat{Y}$ already points to a set of calculated, predicted values or a vector of y values, such as all the `lpsa` or PSA readings included in the `prostate` dataset. Similarly, the term X_j points to the table of all the other variables in the `prostate` dataset. Because there are 8 other variables and close to 100 rows, X is a *matrix*—a higher dimensional table—of values, addressable by coordinates. Finally, β_j are the pivotal coefficients or multiplying quantities that determine the slope or direction of the lines drawn. The second expression equation 3.2 relies more fully on linear algebra. This linear model is written in vector form (Hastie, Tibshirani, and Friedman 2009, 11) or vectorized. The right-hand side comprises two operations X^T, the transpose or rotation of the data, and implicitly—multiplication is hardly ever shown but diagrammed by putting terms alongside each other—an *inner product* of the X matrix and β, the parameters (to use model talk) or coefficients (to use linear algebra talk).[10]

---

think of continuities as subject to ordering. In many ways, that which is continuous bears within it its own ordering, its own immanent seriation or lamination. But in the becoming topological of culture, movement undergoes a transformation according to these authors. Rather than movement as something moving from place to place relatively unchanged (as in geometrical translation), movement should be understood as more like an animation, a set of shape-changing operations. These transformations, I would suggest, should be legible in the way that machine learning, almost the epitome of the processes of modeling and calculation that Lury, Parisi, and Terranova point to, moves through the data. Indeed, the juxtaposition of spam, biomedical data, gene expression data, and handwritten digits already suggests that topological equivalences and a "radical expansion" of comparison might be occurring. Bringing epistopics and topologies together might, I suggest, help trace, map, and, importantly, diagram some of the movements into the data occurring today.

10 Carl Friedrich Gauss and Adrien-Marie Legendre's work on linear regression at this time is well known. The first independent use of linear regression was Gauss' prediction of the location of an "occluded volume," the position of the asteroid Ceres after it reappeared in its orbit behind the sun (Stigler 2002, 320).

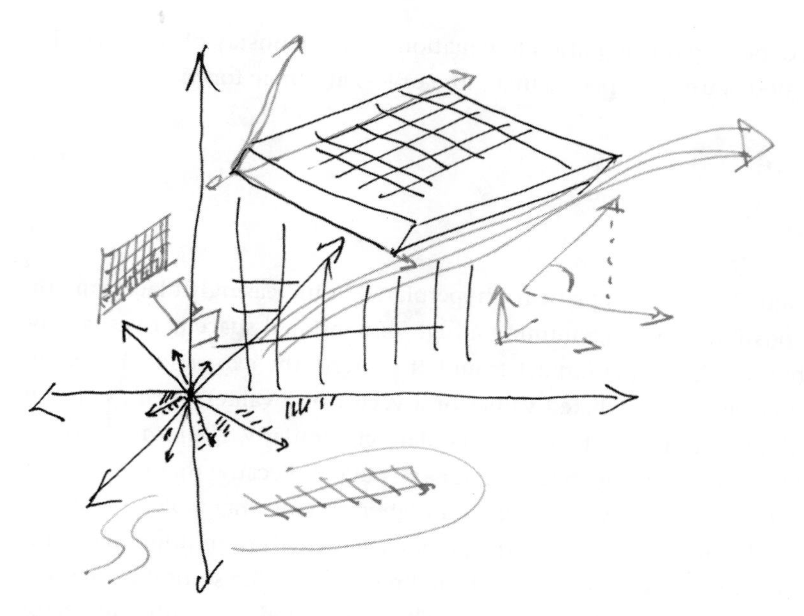

Figure 3.2
Vector space comprises transformations.

The vector form in equation 3.1 diagrams an inclined plane that cannot be fully drawn in any figure, only projected perspectivally onto the surface of a graphic plot. Although that line can never fully come to light, the diagrammatics of equations 3.1 and 3.2 express a way of constructing and orienting it in vector space. Such expressions are epistopic, in that they connect the local complex of activities indexed as tabulated data together through the diagonal diagrammatic element of a line or plane angling through vector space.

Drawing lines in a common space of transformation

Once data are distributed in vector space, machine learners transform that space into the regular forms of the flat loci. Indeed, from the perspective of vector space, machine learners map vector spaces into different ones, usually of lower but sometimes of higher dimensionality. For instance, "drawing" the line of best fit through the `prostate` data or "fitting a line" can be understood as a purely algebraic operation (although in practice most machine learners are not purely algebraic—they optimize and probabilize, as we will see). Viewed in terms of linear algebra, the analytical or closed-form solution for the parameters of the linear model is given in equation 3.3:

$$\hat{\beta} = (\mathbf{X}^T\mathbf{X})^{-1}\mathbf{X}^T\mathbf{y} \tag{3.3}$$

In this expression, linear algebraic operations on the data shown as $\mathbf{X}$ generate the coefficients $\hat{\beta}$ that orient a plane cutting through the vector space.[11] The derivation of the analytical "ordinary least squares" solution relies on differential calculus as well as a range of linear algebra operations, such as matrix transpose, inner product, and matrix inversion, the details of which need not trouble us here. The relevant point is that equation 3.3 constructs a plane—a new vector—that traverses the density shape of a dataset in its full dimensional vector space (nine dimensions in the case of `prostate`).[12]

Implicit vectorization in code and infrastructures

Vectorized transformations of data are the moving substrate of machine learning as it expands, but they are largely taken for granted as given space or commonsense ground. `R` coding practice instantiates vectorization in multiple ways and is sometimes

11 Perhaps more importantly, the linear algebraic expression of these operations presupposes that all the data, both the values used to build the model and the predicted values the model may generate as it is refined or put into operation somewhere, are contained in a common space, the vector space, a space whose formation and transformation can be progressively ramified and reiterated by various lines that either separate volumes in the space or head in a direction that brings along most of the data. Not all of these lines are bound to be straight, and much of the variety and dispersion visible in machine learning techniques come from efforts to construct different kinds of lines or different kinds of decision boundaries (in the case of classification problems) in vector space (e.g., the k-nearest neighbors method does not construct straight lines but somewhat meandering curves that weave between nearby vectors in the vector space; see Hastie, Tibshirani, and Friedman 2009, 14–16). Regardless of whether they are straight, the epistopic aspect of these lines remains prominent. Typically, many different statistical tests (Z-scores or standard errors, F-tests, confidence intervals, and then prediction errors) will be applied to any estimate of the coefficients of even the basic linear regression model well before most advanced or sophisticated models and techniques (cross-validation, bootstrap testing, subset, and shrinkage selection) begin to reconfigure the model in more radical ways.

12 As we will see in the following chapter (chapter 4), it is not always possible to calculate the parameters of a model analytically. Especially in relation to contemporary datasets that have many variables and many instances (rows in the table), linear algebra approaches become unwieldy in their attempt to produce exact results, and machine learning steps in with a variety of computational optimization techniques.

Listing 3.1
Vectorizing code

```
vector1 <- c(0,1,2,3,4,5,6,8,9,10)
vector2 <- c(0,1,2,3,4,5,6,8,9,10)

#procedural programming-style looped addition
result_looped = vector()
for (i in vector1) {
   result_looped[i] = vector1[i] + vector2[i]
}
result_looped

#vectorised addition
result_vectorised  <- vector1 + vector2
result_vectorised
```

described as a "vectorized programming language." The vector space appears and operates just as directly in other programming languages designed for data practice (Octave, Matlab, Python's NumPy, or C++ Armadillo).

In vectorized languages such as R, transformations of a data structure expressed in one line of code simultaneously affect all the elements of the data structure. As the widely used *R Cookbook* puts it, "many functions [in R] operate on entire vectors . . . and return a vector result" (Teetor 2011, 38). Or as *The Art of R Programming: A Tour of Statistical Software Design* by Norman Matloff puts it, "the fundamental data type in R is the *vector*" (Matloff 2011, 24), and indeed in R, all data are vector. There are no individual data types, only varieties of vectors in R. There are many vectorized operations in the R core language and many to be found in packages (the popular plyr package). Vectorized operations can also be found in recent Python data analysis libraries, such as numpy or pandas (McKinney 2012). The fact that many of these vectorized operations occur implicitly suggests how pervasive vector space has become in data practice.[13]

```
[1]  0  2  4  6  8 10 12 16 18 20
[1]  0  2  4  6  8 10 12 16 18 20
```

13 R sometimes presents difficulties for programmers trained to code using so-called procedural programming languages because it so thoroughly embraces the notion of the *vector*—and, hence, regards all data as inhabiting vector space. In many mainstream programming languages, transformations of data rely on loops and array constructs in which some operation is successively repeated on each element of a data structure.

The practical difference between the two approaches to moving through data is illustrated in the code listing 3.1, in which two "vectors" of numbers are added together, in the first case using a classic `for`-loop construct and in the second case using an implicitly vectorised arithmetic operation `+`. The difference between adding $1 \ldots 10$ using a loop or vector arithmetic is completely trivial here, but, as we will soon see, when nested operations are involved, these differences in coding significantly affect human–machine relations. This simultaneity is only apparent because somehow the underlying code has to deal with all the individual elements, but vectorized programming languages take advantage of hardware optimizations or carefully crafted low-level linear algebra libraries.[14] More importantly, this mode of movement is different. Operations no longer step through a series of coordinates that address data elements but instead wield planes, diagonals, cross-sections, and whole-space transformations. Vectorized code reduces both data and computational frictions. The real stake in vectorizing data is not speed but a transformation in data practice. It makes working with data less like iteration through data structures (lists, indexes, arrays, fields, dictionaries, variables) and more like folding a pliable material. Such practical shifts in feeling for data are mundane and yet crucial to the epistopic movements in data.[15]

14 Learning machine learning, and learning to implement machine learning techniques, is largely a matter of implementing a series of matrix multiplications. As Andrew Ng advises his students,

> Almost any programming language you use will have great linear algebra libraries. And they will be high optimized to do that matrix-matrix multiplication very efficiently including taking advantage of any parallelism your computer. So that you can very efficiently make lots of predictions of lots of hypotheses (Ng 2008b, 10:50).

In other parts of his teaching, and indeed throughout the practice exercises and assignments, Ng stresses the value of implementing machine learning techniques for both understanding and using them properly. But this is one case where implementation does not facilitate learning. Ng advises his learners against implementing their own matrix handling code. They should instead use the "great linear algebra libraries" found in "almost any programming language." "Linear algebra libraries" multiply, transpose, decompose, invert, and generally transform matrices and vectors. They will be "highly optimized" not because every programming language has been prepared for the advent of machine learning on a large scale, but rather more likely because matrix operations are just so widely used in image and audio processing. Happily, Ng observes, that means that "you can make lots of predictions" (Ng 2008f). It seems that generating predictions and hypotheses outweighs the value of understanding how things work on this point.

15 A further level of vectorization appears in specific `R` constructs such as `apply`, `sapply`, `tapply`, `lapply`, and `mapply`. All of the `-ply` constructs have a common feature: they take some collection of things (it may be ordered in many different ways—as a list, table, array, etc.), do something to it, and return a collection. Although most programming languages in common use offer

Listing 3.2
Building a prostate model

```
library(ElemStatLearn)
data(prostate)
columns_to_standardize = c(1,2,3,4,6,8,9)
prostate_standard = as.matrix(prostate[, columns_to_standardize])
prostate_standard = as.data.frame(scale(prostate_standard))
prostate_standard = cbind(prostate_standard, gleason=prostate$gleason
    ↪ , svi = prostate$svi, train = prostate$train)
train = prostate$train ==TRUE
prostate_model = lm(lpsa~., prostate_standard[train,-10])
```

Vectorization also motivates increasingly parallel contemporary chip architectures, clusters of computers such as hadoop or spark, reallocation of computation to Graphic Processing Units (GPUs), data-center usage of Field Programmable Gate Arrays (FPGAs), and various other Cyclopean infrastructures of cloud computing. Many of these condensing and expanding movements of data are diagrammed in miniature in the R constructs as operators in vector space.

Lines traversing behind the light

How does the combination of algebraic vector space and vectorized code play out in data? "We fit a linear model," write Hastie and co-authors, referring to the leading epistopic operation on data in *Elements of Statistical Learning*. In R this might look like the code excerpt shown in listing 3.2.

Table 3.3 displays estimates of the coefficients or parameters $\hat{\beta}$ that define the direction of a flat surface running through the vector space of the prostate data.[16]

constructs to help deal with collections of things sequentially (e.g., by accessing each element of a list in turn and doing something with it), R offers ways of expressing a simultaneous operation on them all. The -ply constructs ultimately derive from the functional logic developed by the mathematician Alonzo Church in the 1930s (Church 1936, 1996). The functional programming style of applying functions to functions seems strangely abstract.

16 From the epistopic viewpoint, the most obvious result of fitting a linear model is the production not of a line on a diagram or in a graphic. As we have seen, such lines cannot be easily rendered visible. Instead, the model generates a new column-vector of coefficients (see table 3.3) and some new numbers, *statistics*. This table is not as extensive as the original data, the X and *Y*

Table 3.3
Fitting a linear model to the prostate dataset

Estimate	Std. Error	t value	Pr(>|t|)
0.0227	1.1750	0.02	0.9847
0.5887	0.1097	5.37	0.0000
0.2279	0.0828	2.75	0.0079
−0.1226	0.0878	−1.40	0.1681
0.1821	0.0886	2.06	0.0443
−0.2499	0.1339	−1.87	0.0670
0.2313	0.1331	1.74	0.0875
−0.0256	0.1742	−0.15	0.8839
0.6386	0.2586	2.47	0.0165

This new vector is a product of operations in the vectorized `prostate` data. Some vectorizing operations can be seen in `R` code in listing 3.2 (e.g., `as.matrix` or `scale(prostate_standard)`).

The unique solution to the problem of fitting a linear model to a given dataset using the popular method of least squares (Hastie, Tibshirani, and Friedman 2009, 12) is given by the operations we have seen in equation 3.3. This tightly coiled expression calculates the $\hat{\beta}$ parameters that set the slope and location of a flat surface or plane in nine-dimensional vector space using all of the `prostate` variables apart from one variable chosen as the response or predicted variable, in this case, `lpsa`. X and y matrices are multiplied, transposed (a form of rotation that swaps rows for columns), and inverted (a more complex operation that finds another matrix) in a series of linear algebra transformations. Epitomizing the implicitly vectorized code often seen in machine

vectors. But the names of the variables in the dataset appear as rows in the new table, a table that describes something of how a line has been fitted by the linear model to the data. The columns of the table now bear abbreviated and much more statistical names such as `estimate` (the estimated values of $\hat{\beta}$, the key parameters in any linear model), `Std. Error`, `t value`, and the all-important p values written as `Pr(|t|)`. The numerical values ranging along the rows mostly range from −1 to 1, but the final column includes values that are incredibly small: `1.47e-06` is a few millionths. Other statistics range around the outside the table: the `F-statistic`, the `R-squared` statistic, and the `Residual standard error`. The numbers of the table 3.2 become epistopic here because they now appear as a set of standard errors, estimates, t-statistics, and p values that together indicate how likely the estimated values of β are and therefore how well the diagonal line expresses the relations between different dimensions of the dataset in the vector space.

Listing 3.3
Closed-form evaluation of linear model parameters

```
beta_hat = ginv(t(X) %*% X) %*% t(X) %*% y
```

learning, calculating $\hat{\beta}$ for the `prostate` data only requires one line of `R` code, as shown in listing 3.3.

The implicit vectorization of the `R` code in the code listing 3.3, the fact that it already concretely operates in the vector space, operationalizes the diagram of equation 3.3 as a machine process. The vectorized multiplication, transposition, and inversion of data creates the new vector $\hat{\beta}$ whose variations can be explored, observed, graphed, and varied in ways that go well beyond the statistical tests of significance, variation, and error reported in table 3.3. (We will have occasion to return to these statistical estimates in chapter 5.) The play of values that starts to appear even in fitting one linear model will become much more significant when fitting hundreds or thousands of models, as some machine learners do.[17]

The vectorized table?

Vectorization and the transforming movements it projects and proliferates bring us back to the problem of how machine learners mix datasets that span different settings. In short, vectorizing computation produces the vector space, which we might understand as a resurgent form of the highly associative pre-Classical table, a table that tolerates many relations and similarities, operationally concrete and machinically

17 This differentiation is important: it is not typical machine learning practice to construct one model, characterized by a single set of statistics (F scores, R^2 scores, *t* values, etc.). In practice, most machine learning techniques construct many models, and the efficacy of some predictive techniques derives often from the multiplication or indeed proliferation of models. Techniques such as neural networks, cross-validation, bagging, shrinkage and subset selection, and random forests, to name a few, generate many statistics, and navigating the multiple or highly variable models that result becomes a major concern. An epistopic abundance will appear here—bias, variance, precision, recall, training error, test error, expectation, Bayesian Information Criteria, and so on, as well as graphisms such as Receiver-Operator-Characteristics (ROC) curves. Put simply, the proliferation of models starts to drive the dimensional expansion of the vector space. At the same time, the multiplicity of models multiplied by the machine learners becomes the topic of statistical analysis.

abstract. It is no longer a visible diagram but a machinic process that multiplies and propagates into the world along many diagonal lines.

Machine learning relies on a broad but subtle transformation of data into vectors and a vector space. Slightly repurposing Foucault's archaeology of tables in *The Order of Things*, I have suggested that vectorization remaps the grid of the table into the expanding dimensions of the vector space. This space accommodates both simple and complex natures. This is not the first such expansion of the table. We need only think of the relational database systems of the late 1960s and their multiplication of tables (Mackenzie 2012). But in the vectorized and matrix-form practices of the vector space, machine learning produces for the first time a meta(s)table volume that cannot be surfaced on a page or screen.

Does the vectorization of data lie a "a long way from sense data" as Arendt suggests? In the diagrammatic operations of linear algebra on data and the vectorization of code, machine learning traverses dimensions that, as Arendt observes, cannot be immediately sensed. Whitehead's notion of data strain as "a complex distribution of geometrical significance" suggests, however, that vectorization is not a complete loss of feeling. Every machine learner inhabits and moves through the vector space along different strains. Sometimes their operations flatten the vector space down into lower dimensional subspaces, as in the many "dimensional reduction" machine learners such as principal component analysis, Latent Dirichlet Allocation, or indeed the linear regression model that maps an irregular volume onto a plane. Sometimes they expand the vector space into a great many new dimensions or features (as we saw with "dummy variables" that embody categories and as we will see with support vector machine classifiers in chapter 6 or the deep learners of chapter 8).

The epistopic transformation of datasets and tables into vector space reaches into and realigns norms, regularities, associations, and classifications. It acts as a powerful tensor on knowledges and operations of many different kinds as it transposes, inverts, and remaps local complexes of activity. In following what happens to vectors, lists, matrices, arrays, dictionaries, sets, dataframes, series, or tuples in data, we might get a sense of how the epistemic operations of predictive models, supervised and unsupervised learners, classifiers, decision trees, and neural networks have purchase in data.

What is at stake in vectorizing data? It produces a common space that juxtaposes and mixes complex localized realities. The `prostate` dataset could be aggregated and melded as vectors with a microarray, heart disease, or bone density datasets. In vector space, identities and differences change in nature. Similarity and belonging no longer rely on resemblance or a common genesis but on measures of proximity or distance, on flat loci that run as vectors through the space. Vectorization, the deep saturation of the table by algebra, constitutes all relations as movements of transformation, diagonalization, inversion, or rotation. The epistopic power of vectorization takes root in the elementary practices of machine learning and engenders many variations among machine learners. Vectorization also strains the production of knowledge through a loss of the visible geometry of tabular comparison. This loss of visibility is, as we will see, was met by the production of new groups of statements, visible forms and operational devices, and infrastructures that accommodate the dimensional expansion of vector space.

The fascination of machine learning, its seemingly endless applications (I refer the reader back to the diagram of machine learning's vastness in chapter 2), owes much to the vector feeling, with its twin lures of ideal operationality—everything is a vector operation—and its tantalizing tendency to expand and move. This feeling, the vector feeling, we might note, is not surprising. "Characteristically" for Whitehead, "feelings are 'vectors'; for they feel what is there and transform it into what is here" (Whitehead 1960, 87).

Expansive data vectorization challenges contemporary critical thought to develop intuitions and value-relevant concepts describing vector-feelings or data strains. We lack good intuitions of how to do that partly because machine learning implicitly vectorizes its practice in code, infrastructures, and highly condensed diagrammatic forms. Archaeology of the transformations of tables into vector spaces seeks to unwind or de-diagonalize some of the operations rippling through different treatments of data. The act of diagramming how machine learners vectorize data densities begins to locate and unravel the processes of knowing, predicting, and deciding on which many aspects of the turn to data rely. The vectoral operations we have just been viewing are organized and aligned by other lines of diagrammatic movement that shape surfaces in more convoluted forms.

4 Machines Finding Functions

Because of a gradient that no doubt characterizes our cultures, discursive formations are constantly becoming epistemologized (Foucault 1972, 195).

"All knowledge," hypothesizes Pedro Domingos, "past, present,and future can be derived from data by a single, universal learning algorithm" (Domingos 2015, 25). How will the "single, universal" algorithm learn, how will it "epistemologize," to use Foucault's term, "our cultures"?

In practice, the opening pages of machine learning textbooks often warn or enthuse about the profusion of techniques, algorithms, tools, and machines. "The first problem facing you," cautions Domingos' readers of the *Communications of the ACM*, "is the bewildering variety of learning algorithms available. Which one to use? There are literally thousands available, and hundreds more are published each year" (Domingos 2012, 1). "The literature on machine learning is vast, as is the overlap with the relevant areas of statistics and engineering," echoes David Barber in *Bayesian Reasoning and Machine Learning* (Barber 2011, 4); "statistical learning refers to a vast set of tools for understanding data," writes James and coauthors in an *Introduction to Statistical Learning with R* (James et al. 2013, 1); or writing in *Statistical Learning for Biomedical Data*, the biostatisticians James Malley, Karen Malley, and Sinisa Pajevic "freely admit that many machines studied in this text are somewhat mysterious, though powerful engines" (Malley, Malley, and Pajevic 2011, 257). In *Thoughtful Machine Learning*, Matthew Kirk exacerbates the situation: "flexibility is also what makes machine learning daunting. It can solve many problems, but how do we know whether we're solving the right problem, or actually solving it in the first place?" (Kirk 2014, ix). The prefatory comments from Domingos, Barber, James, Malley, and Kirk suggest a rampant even weed-like abundance of machine learners, as does the 700 or so pages of *Elements of Statistical Learning*. Much learning of machine learning work, at least for machine learners, concerns not so much implementation of particular techniques (neural network, decision tree, support vector machine, logistic regression, etc.) but rather navigating the maze of

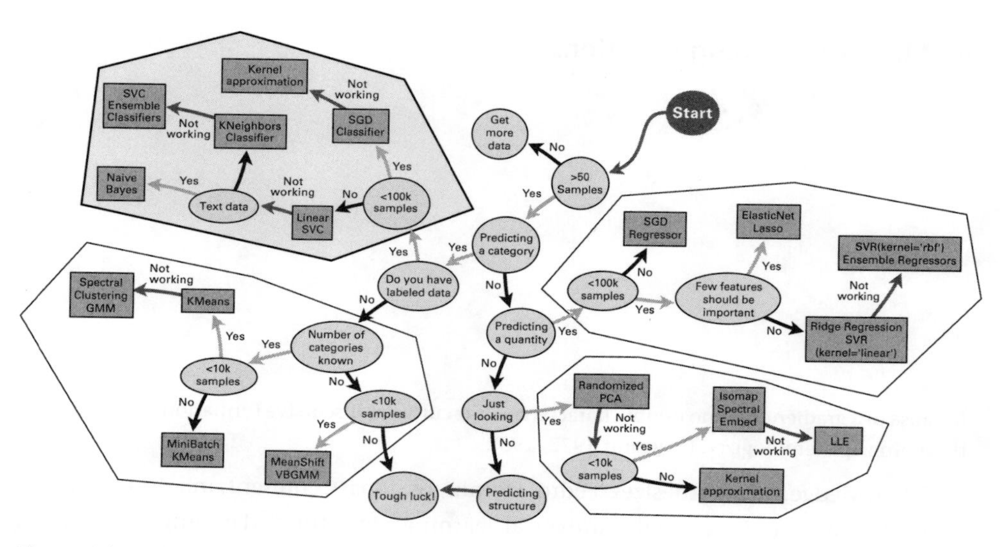

Figure 4.1
`Scikit-learn` map of machine learning techniques.

methods and variations that might be relevant to a particular situation. How does this dual effect of profuse accumulation and the ideal a single, universal machine learner arise and hold together?

The machine learners I have just cited present that profusion as a problem of the piling up of techniques. As the authors of textbooks and how-to manuals, they attempt to manage it by providing, indexes, maps, and guides to the bewildering variety of machine learners. *Elements of Statistical Learning* deploys tables, overviews, theories of statistical modeling, model assessment, and comparison techniques to aid in navigating them.

Parallel and complementary mappings accompany software libraries. The visual map of machine learning techniques shown in figure 4.1 comes from a machine learning library written in `Python`, `scikit-learn` (Pedregosa et al. 2011). This software library is widely used in industry, research, and commerce. In contrast to the pedagogical expositions, theoretical accounts or guides to reference implementation, or the many overlapping packages in `R`, code libraries such as `scikit-learn` order the range of techniques by offering recipes and maps for the use of the *functions* the libraries supply. The branches in the figure lay down paths through the profusion of techniques as a decision tree.[1]

1 Similarly, for `R` code, the *Comprehensive R Archive Network* tabulates key libraries of `R` code in a machine learning "task view" (Hothorn 2014).

The architecture of software libraries classifies and orders machine learners. `Scikit-learn`, for instance, comprises a number of subpackages. Modules such as `lda` (linear discriminant analysis), `svm` (support vector machine), or `neighbors` (*k* nearest neighbors) point to well-known machine learners, whereas `cross-validation` or `feature_selection` refer to ways of testing models or transforming data, respectively. These divisions, maps, and classifications help order the techniques, but they obscure the process that first generates a competing profusion of machine learners.

If, as I have suggested earlier, we understand knowledge in terms of the radically reconceptualized operational statements that Foucault described in *The Archaeology of Knowledge*, then statements comprise various units (sentences, series, tables, propositions, diagrams, equations, numbers) mapped to a field of objects, subject positions, and domains of coordinations and reuse by an *enunciative function* (Foucault 1972, 106). Confronted by a profusion of machine learners and the idea of a single, universal machine learning, an archaeological analysis attends to the enunciative function that multiplies meanings and operations.

We might understand the enunciative function as the generative process that proliferates machine learners. The listing and mapping of accumulated techniques, whether in the form of textbooks such as *Elements of Statistical Learning* or a code library such as `scikit-learn`, together with the many attempts to unify them (Domingo's "single, universal algorithm," `scikit-learn`'s map, *Elements of Statistical Learning*'s statistical theory), suggests a commonality in the production of statements. As I will argue in this chapter, there are many techniques, algorithms, and ways of deriving knowledge from data in machine learning because statements are actually rare in this operational formation. "Because statements are rare," writes Foucault, "they are collected in unifying totalities, and the meanings to be found in them are multiplied" (Foucault 1972, 120).

Learning functions

The rarity of statements amid the profusion of machine learners revolve around a single operator, the function.[2] Table 4.1 shows the titles and author-supplied keywords of a sample of well-cited machine learning publications. In these randomly chosen publications, mathematical functions—"kernel function," "discriminant function," "radial basis function"—mingle with biological and engineering functions—"protein-binding

2 I discuss the sense of function as operation or process in chapter 7. There I suggest that this important sense of function as operation or process, a sense that has underpinned transformations in life, social, and clinical sciences, may be shifting toward a different ordering.

Table 4.1

Sample of highly cited machine learning publications referring to "function" in title or keyword

Year	Title	Keywords	Citations
1995	Recurrent Radial Basis Function Networks For Adaptive Noise Cancellation	radial basis function network; noise cancellation; nonlinear filtering	65
1997	Comparing Support Vector Machines With Gaussian Kernels To Radial Basis Function Classifiers	clustering; pattern recognition; prototypes; radial basis function networks; support vector machines	392
1999	Bump Hunting In High Dimensional Data	data mining; noisy function optimization; classification; association; rule induction	103
2000	Block Coordinate Relaxation Methods For Nonparametric Wavelet Denoising	basis pursuit; block coordinate relaxation; function estimation; interior-point; optimization	90
2001	Efficient Source Adaptivity In Independent Component Analysis	blind signal separation; independent component analysis (ica); score function estimation; source adaptivity	40
2003	Hierarchical Learning Architecture With Automatic Feature Selection For Multiclass Protein Fold Classification	feature extraction; gating network; n-gram coding; protein sequence; radial basis function network (rbfn); structure classification of protein (scop); support vector machine (svm)	34
2005	Sensitivity Analysis Applied To The Construction Of Radial Basis Function Networks	sensitivity analysis; radial basis function neural network; orthogonal least square learning; network pruning	38
2005	Predicting Enzyme Class From Protein Structure Without Alignments	protein function prediction; structure; ec number; machine learning; structural genomics	78
2005	Partially Supervised Classification Of Remote Sensing Images Through Svm Based Probability Density Estimation	probability density function (pdf) estimation; supervised classification; support vector machines (svms); unknown classes	39
2006	Structural Bioinformatics Prediction Of Membrane Binding Proteins	protein-membrane interactions; function annotation; support vector machines; peripheral proteins; protein function prediction	29
2006	Fully 3 D Pet Reconstruction With System Matrix Derived From Point Source Measurements	positron emission tomography (pet) iterative reconstruction; point spread function (psf) modeling; system response	151

2006	Automated Discovery Of 3d Motifs For Protein Function Annotation		43
2007	2d Rna Coupling Numbers: A New Computational Chemistry Approach To Link Secondary Structure Topology With Biological Function	rna secondary structure; molecular descriptors; sequence-function relationships; coupling numbers; linear classifiers; machine learning algorithms	52
2007	Choosing Parameters Of Kernel Subspace Lda For Recognition Of Face Images Under Pose And Illumination Variations	gaussian radial basis function (rbf) kernel; generalization capability; kernel fisher discriminant (kfd); kernel parameter; model selection	33
2007	A Balanced Accuracy Function For Epistasis Modeling In Imbalanced Datasets Using Multifactor Dimensionality Reduction		70
2008	Identification Of Catalytic Residues From Protein Structure Using Support Vector Machine With Sequence And Structural Features	active site; protein function prediction; functional residues; sequence-structural features; spatial neighbors	23
2008	Comparative Qsar Studies Of Cyp1a2 Inhibitor Flavonoids Using 2d And 3d Descriptors	2d and 3d descriptors; cyp1a2; genetic function approximation; genetic partial least squares; partial least square; qsar	35
2011	Quat 2l: A Web Server For Predicting Protein Quaternary Structural Attributes	smart; function domain composition; pseudo amino acid composition; complexity measure factor; fuzzy k nearest neighbor	30
2013	Classification Of Emg Signals Using Pso Optimized Svm For Diagnosis Of Neuromuscular Disorders	electromyography (emg); motor unit action potentials (muaps); discrete wavelet transform (dwt); radial basis function networks (rbfn); k-nearest neigbour (*k*-nn); particle swarm optimization (pso); support vector machine (svm); parameter selection	30

function," "intestinal motor function," or "rules to control locomotion." Mathematical functions, however, dominate. Machine learners "find," "estimate," "approximate," "analyse," and sometimes "decompose" mathematical functions. The primary mathematical sense of a function refers to a relation—a mapping—between sets of values or variables. (A variable is a symbol that can stand for a set of numbers or other values.) A function is a one-to-one relation between two sets of values. It maps a set of arguments (inputs) to a set of values (outputs, or to use slightly more technical language, it maps between a *domain* and a *co-domain*). As we have already seen, mathematical functions are often written in formulae of varying degrees of complexity. They are of various genres, provenances, textures, and shapes: polynomial functions, trigonometric functions, exponential functions, differential equations, series functions, algebraic or topological functions, and so on. Various fields of mathematics have pursued the invention of functions. In machine learning and information retrieval, important functions include the logistic function (discussed later), probability density functions (PDFs) for different probability distributions (Gaussian, Bernoulli, Binomial, Beta, Gamma, etc.; see chapter 5), and error, cost, loss, or objective functions (these four are almost synonymous). (The latter group I discuss below because they underpin many of the claims that machine learners learn.)

As we will see, from the perspective of the function, machine learning can be understood as a function-finding operation. Implicitly or explicitly, machine learners find a mathematical expression—a function—approximating the social, technical, financial, transactional, biological, brain, heart, or group process that flowed the data in question into vector space. Regardless of the application, no single mathematical function perfectly or uniques expresses data. Many, if not infinite, functions can approximate any given data. Even if there were a master algorithm, therefore, it would be concerned with a field of functions, and it would entail observation, classification, and selection (finding, in short) in deriving knowledge from data.

Supervised, unsupervised, and reinforcement learning and functions

The capacity of machine learners to learn is closely linked to forms of observation that accompany and orient it. The optics of this observation of machine learners vary, but they are always partial or incomplete partly because of the dimensionality of vector space and partly because of the domains in which machine learning operates. Although the field is pragmatic in its commitment to classification and prediction (and curiously idealistic too in its constant reuse of well-worked datasets such as `iris` or `South African heart disease`), it distinguishes among three broadly different kinds of *learning*—supervised, unsupervised, and reinforcement—in terms of their observability. *Elements of Statistical Learning* presents the distinction between supervised and unsupervised learning:

> With supervised learning there is a clear measure of success or lack thereof, that can be used to judge adequacy in particular situations and to compare the effectiveness of different methods over various situations. Lack of success is directly measured by expected loss over the joint distribution $Pr(X, Y)$. This can be estimated in a variety of ways including cross-validation. In the context of unsupervised learning, there is no such direct measure of success. . . . This uncomfortable situation has led to heavy proliferation of proposed methods, since effectiveness is a matter of opinion and cannot be verified directly (Hastie, Tibshirani, and Friedman 2009, 486–487).

Supervised learning in general terms constructs a model by training on some sample data (the training data) and then evaluating the model's effectiveness in classifying or predicting unseen test data whose actual values are already known. The "clear measure of success" in relation to so-called "supervised learning" is of relatively recent date.[3] Unsupervised machine learning techniques generally look for a range of well-characterized patterns in the data without any training or testing phases (e.g., *k*-means or principal component analysis do this, and both techniques have been heavily used for more than 50 years). In both supervised and unsupervised learning, machine learners observe how a function (or functions) changes as a model transforms, partitions, or maps the data.

Viewed as an enunciative function, machine learning makes statements through operations that treat functions as partial observers. At the same time, opacity—"no direct measure of success"—is generative in machine learning. *Elements of Statistical Learning* admits that success cannot be measured and that this inaccessibility has led to a proliferation of methods, transformations, and changes. If, as the first part of the quoted text puts it, supervised learning has a clear "measure of success," then that success only seems to encourage further variations and comparisons that end up proliferating machine learners, their publications, and their software implementations.

Which function operates?

Differences between machine learners can be described using mathematical functions. That is, machine learners operate as functions, and observations of those operations also constitute functions. Functions instigate *both* the operations and ordering of those operations.

3 Only in the mid-1980s were the first theories of algorithmic learning formalized (Valiant 1984).

For instance, classifiers, or machine learners that classify, are often identified directly with functions:

A classifier ... is a function $d(\mathbf{x})$ defined on $\mathcal{X}$ so that for every $\mathbf{x}$, $d(\mathbf{x})$ is equal to one of the numbers $1, 2, \ldots, J$ (Breiman et al. 1984, 4).

Writing in the 1980s, the statistician Leo Breiman identifies classifiers—perhaps the key operational achievement of machine learning and certainly the catalyst of many applications—with functions. A classifier *is* a function $d(\mathbf{x})$, where $\mathbf{x}$ is the data and d ranges over numbers that map onto categories, rankings, or other forms of order and belonging (e.g., `cat` or `not cat` in the case of `kittydar`).

The identification of machine learning with functions appears in the first pages of most machine learning textbooks. Viewed operationally, learning in machine learning means finding a function that can identify or predict patterns in the data. As *Elements of Statistical Learning* formulates it,

our goal is to find a useful approximation $\hat{f}(x)$ to the function $f(x)$ that underlies the predictive relationship between input and output (Hastie, Tibshirani, and Friedman 2009, 28).

This statement of learning laminates several layers into the function. It posits the existence of *the* function that generated the data as a foundation. This function figures as a ground truth existentially imputed to the world. It also refers to "finding ... $\hat{f}(x)$," where the '^' indicates a useful approximation. A leading machine learning theorist, Vladimir Vapnik echoes the statement of learning as approximation: "learning is a problem of *function estimation* on the basis of empirical data" (Vapnik 1999, 291).[4] The use of the term "learning" in machine learning displays affiliations to the field of artificial intelligence, but the "function-fitting paradigm" as Hastie, Tibshirani, and Friedman (2009, 29) term it, emphasizes this double layering of function as an observed approximation. Most importantly, *learning* here is understood as finding. Despite many differences in the framing of the techniques, all accounts of machine learning, even those such as *Machine Learning for Hackers* (Conway and White 2012) that eschew any explicit recourse to mathematical formula, rely on the formalism and modes of thought associated with mathematical functions. Whether they are seen as forms of artificial intelligence or statistical models, machine learners are directed to build "a good and useful approximation to the desired output" (Alpaydin 2010, 41), or, put more statistically, "to use the sample to find the function from the set of admissable functions that minimizes the probability of error" (Vapnik 1999, 31).

4 Vapnik is said to have invented the support vector machine, one of the most heavily used machine learning technique of recent years, on the basis of his theory of computational learning. Chapter 6 discusses the support vector machine.

The superimposed or doubling of function as operation and observer is hardly ever explicitly mentioned by machine learners. The pages of Hastie, Tibshirani, and Friedman (2009) present a litany of functions: quadratic function, likelihood function, sigmoid function, loss function, regression function, basis function, activation function, penalty functions, additive functions, kernel functions, step function, error function, constraint function, discriminant function, probability density function, weight function, coordinate function, neighborhood function, and the list goes on. This mixed list draws from a pool of several hundred mathematical functions commonly used in science and engineering.[5] Clearly neither machine learners nor critical researchers can expect to understand the functioning of all these functions in any great detail. Although this prickly list of terms confirms the salience of functions in machine learning (as perhaps in many other science and engineering disciplines), certain basic differences between them might be a way to map the interplay of operational and observational functions. We can already see in this list that functions are diverse. Sometimes the function refers to a mathematical form—quadratic, coordinate, basis, or kernel; sometimes it refers to statistical considerations—likelihood, regression, error, or probability density; and sometimes it refers to some other concern that might relate to a particular modelling device or diagram—activation, weight, loss, constraint, or discriminant.

What does a function learn?

We wish to know: in what sense does a machine learner learn? This question can now be reframed: how do machine learners find functions? For critical thought, this question is vexing because if function-finding agency inheres in machines and devices, then the politics of human–machine relations and the practices of knowledge production shift. The philosopher of science Isabelle Stengers sets tight limits on functions:

> No function can deal with learning, producing, or empowering new habits, as all require and achieve the production of different worlds, non-consensual worlds, actively diverging worlds (Stengers 2005, 162).

If they cannot learn new habits, then what can functions learn? In some ways, Stengers would, on this reading, be taking a fairly conventional position on

5 The U.S. National Institute of Standards published *The Handbook of Mathematical Functions* in 1965 (Abramowitz 1965). This heavily cited volume, now also versioned online, lists hundreds of functions organized in various categories, ranging from algebra to zeta functions. Although a number of the functions and operations catalogued there surface in machine learning, machine learners implement, as we will see, quite a narrow range of functions.

mathematical functions. They cannot learn or produce anything, only reproduce patterns implicit in their structure. Similar statements might be found in many philosophical writings on science and mathematics in particular.[6] But throughout her writing, Stengers explicitly affirms *experimental practice*, much of which depends on functions and their operations (Stengers 2008). It might be better to say that she limits the agency of functions in isolation to highlight their specific power in science: "celebrating the exceptional character of the experimental achievement very effectively limits the claims made in the name of science" (Stengers 2011, 376). (Limiting claims made for science might save it from being totally repurposed as a technoeconomic innovation system.)

The connection between a given function and a given concrete experimental situation is highly contingent or indeed singular. Stengers argues that mathematical functions impinge on matters of fact via experimentally constructed relays:

> The reference of a mathematical function to an experimental matter of fact is neither some kind of right belonging to scientific reason nor is it an enigma, but actually the very meaning of an experimental achievement (Stengers 2005, 157).

The generic term "reference" here harbors a multitude of relations. The experimental achievement, the distinctive power of science, works through a tissue of relations that connect people, things, devices, facts (statements), and mathematical functions in a heterogeneous weave.[7] Given that learning is not radically innate to machines, it might be better understood as an experimental achievement. When a biomedical researcher seeks to "estimate the probability that a critically-ill lupus patient will not survive the first 72 hours of an initial emergency hospital visit"

6 A major reference here would be Ernst Cassirer (Cassirer 1923), who posited a philosophical-historical shift from ontologies of substance reaching back to Aristotle's categories (Aristotle 1975) to a functional ontology emerging in the 19th century as the notion of function was generalized across many mathematical and scientific fields. (See Heis 2014 for a recent account of the *Funktion-Begriff* in Cassirer's philosophy.) In a recent article, Paolo Totaro and Domenico Ninno suggest that the transition from substance to function occurs practically in the form of the algorithm (Totaro and Ninno 2014). The idea of computable functions lies at the base of theoretical computer science and has been a topic of interest in some social and cultural theory (e.g., (Parisi 2013); see also (Mackenzie 1997)), but Totaro and Ninno suggest that algorithmic processes, as the social practice of the function, form contradictory hybrids with remnants of substance, in particular, categories and classification. They see bureaucratic logic, for instance, as hopelessly vitiated by a contradiction between classification and function. Machine learning, I'd suggest, is an important counterexample. It hybridizes function and classification without any obvious contradiction.

7 This point has often been made in the social studies of science (see Latour 1993 for a high-level account).

(Malley, Malley, and Pajevic 2011, 5), he or she might estimate and evaluate his or her predictions using classical statistical approaches (analysis of variance, correlations, regression analysis, etc.). The question from Stengers' standpoint is this: what happens to the structure of referrals through experiments and the existing knowledge when functions are said to learn? To address this question, we need to delineate how functions function in machine learning.

At first glance, machine learning as a field is not experimental (even if it radically influences the conduct of experiments in many scientific fields; see chapter 7). It lacks the apparatus, instruments, laboratories, field sites, or clinics of experimental practice. Experimentation takes place principally in the form of rendering diagrammatically the relays or referrals between different functions as they traverse data. They appear in graphic forms as plots. The diagrammatic entanglement of operation and observation in functions is not surprising. The historical invention of the term "function" and a notation for writing functions by the philosopher G.W. Leibniz in the 17th century pertained to the problem of describing continuous variations in curves. Functions for Leibniz describe variations in response to other changes (y may change in response to a change in x), but they can also describe tendencies in functions (as a derivative function describes the sensitivity or rate of change of the slope of curve). Identifying and locating important tendencies or changes in functions—*singularities* in curves—also preoccupies the function-finding done in machine learning. In contrast to the vector space that expands to accommodate all transformations, the many observational elements such as graphic objects permit observations of tendencies or change points. The experimental relay or referral, the power to confer on things the power to confer on the machine learner (human–machine) the power to speaker in their name (Stengers 2000, 89), pivots around the double layer of functions. An operational function transforms the vector space and an observational function generates statements concerning degrees and rates of success, fit, or error.

Although we have yet to see how a function can observe, we can readily see some of the effect of the coupling between operational and observational functions. In the several hundred color graphic plots in *Elements of Statistical Learning*, a striking mixture of network diagrams, scatterplots, barcharts, histograms, heatmaps, boxplots, maps, contour plots, dendrograms, and 3D plots exhibit different aspects of this tension between operation and observation. Many of these graphic forms are common in statistics as statements of variation or tendency in data (histograms and boxplots). Others relate specifically to machine learning (e.g., Receiver Operating Curve [ROC] or regularization path plots). A significant proportion of these graphics do not display data from experiments or measurements, but diagram variations in the operational function that transforms the data in relation to some criteria of observations (e.g., prediction errors or purity of classification).

Observing with curves: The logistic function

How can a function observe? As we have already seen, machine learners often learn by fitting, as well as over-fitting and under-fitting. *Fitting* is a way of bringing functions into the data. As we saw in the previous chapter, the vector space cannot be fully seen, and its operational transformation into lines, planes, or smooth surfaces often remains occluded. Graphic plots and statistical summaries offer perspectival views on those transformations, but machine learners observe many of those transformations by adding a feedback loop between the transformations (fitting a line, building a decision tree, adjusting the weights in a neural net, etc.) and observed outcome.

Take the example of *sigmoid* functions. These quite simple functions underpin many classifiers and animate many of the operations of neural network, including their recent reincarnations in deep learning (Hinton and Salakhutdinov 2006; Mohamed et al. 2011). As operational functions in machine learning, they illustrate a transformation of continuous values into discrete values. As observational functions, they exemplify observability, as we will see, in the form of their differentiability. An example of a sigmoid function, the logistic function, can be written as:

$$f(x) = 1/(1 + e^{-kx}) \tag{4.1}$$

The logistic function (shown as equation 4.1 and as two curves in figure 4.2), as we will see, is very important in many classification and decision settings partly due to the *non-linear* shape that constrains vertical movement within the values (0 to 1), and partly because of the range of shapes opened up by the parameter K. How does a function such as the sigmoid function 'observe' anything? Here the S-shape of the curve itself and even the name 'sigmoid' is the best guide. The logistic function has quite a long history in statistics since that curve diagrams growth and change in various ways. As the historian of statistics J.S. Cramer writes: 'The logistic function was invented in the nineteenth century for the description of the growth of organisms and populations and for the course of autocatalytic chemical reactions' (Cramer 2004, 614).[8] In nearly

8 The Belgian mathematician Pierre-François Verhulst designated the sigmoid function the 'logistic curve' in the 1830–40s (Cramer 2004, 616). It was independently designated the 'autocatalytic function' by the German chemist Wilhelm Ostwald in the 1880s, and then re-invented under various names by biologists, physiologists and demographers during 1900–1930s (617). The term

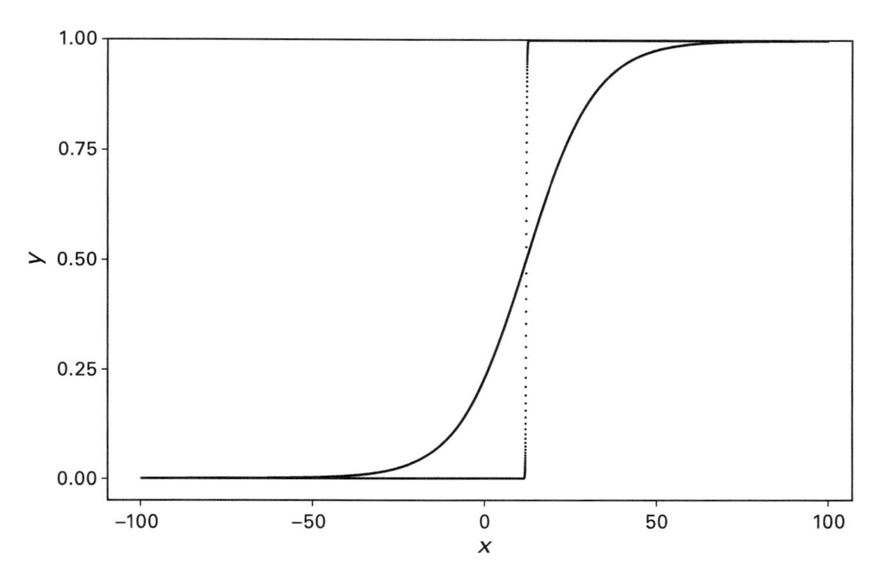

Figure 4.2
Logistic or sigmoid function for two different values of the parameter *k*.

all of these cases, the function was used to fit a curve to data on the growth of something: populations, reactions, tumours, tadpoles tails, oats and embryos. The reference of the curve to growth comes from its changing slope. Growth starts slowly, increases rapidly, and then slows down again as it reaches a limit. In the second half of the 20th century, it was widely used in economics. Census data and clinical or laboratory measurements supplied the actual values of $f(x)$ at particular times, the x values. The task of the demographer, physiologist, or economist was to calculate the values of parameters such as k that controlled the shape of the curve.

Historically, then, the logistic function has a well-established biopolitical resonance. But note that the curves showing in figure 4.2 plot the same data (X and y values) yet differ in their curvature. This diagrammatic variation derives from the parameter k, which discreetly appears in the equation 4.1 next to x. Such parameters are vital control points in function fitting and any learning associated with that. Varying these parameters and optimizing their values is the basis of "useful approximation" in machine learning.

'logistic' returns to visibility in the 1920s, and has continued in use as a way of describing the growth of something that reaches a limit.

Sometimes these parameters can be varied so much as to suggest entirely different functions. In equation 4.2, for instance, $k = 12$ produces a much sharper curve, a curve that actually looks more like a qualitative change range than a smooth transition from 0 to 1. The sharp shape of the logistic curve when the scaling parameter k is larger transforms the function into a classifier, into a function that, as Breiman puts it, is equal to one of the numbers 0 or 1. In this setting, the function $f(x) = 1/(1 + e^{-x})$ maps continuously varying numbers (the x values) onto a domain of discrete values. Because $f(x)$ tends very quickly to converge on values of 1 or 0, it can be coded as no, survived/deceased, or any other binary difference. The mapping between the x values slides continuously and the binary difference pivots on the combination of the exponential function (e^{-x}), which rapidly tends toward zero as x increases and rapidly tends toward ∞ as x decreases, and the dividend $\frac{1}{1+\cdots}$, which converts high-value denominators to almost zero and low-value denominators to one. This mapping between variations in x and the value of the function $f(x)$ is mathematically elementary but typical of the relaying of references that allows functions to intersect with and constitute matters of fact and states of affairs such as `cat` and `not-cat`. This realization—that a continuously varying sigmoid function can map discrete outcomes—forms the basis of many machine learning classifiers.

The cost of curves in machine learning

I have been suggesting that experimentality in machine learning consists of coupling operational and observational functions. If operational functions move through or transform the data, observational functions render the effects of those transformations visible. How does this take place practically? The logistic function appears frequently in machine learning literature, prominently as part of perhaps the most vernacular machine learner, the logistic regression model (see table 4.2 for a sample of well-cited publications). Descriptions of logistic regression models appear in nearly all machine learning tutorials, textbooks, and training courses (see chapter 4 in Hastie, Tibshirani, and Friedman 2009). In biomedical research, "logistic regression is the default 'simple' model for predicting a subject's group status" (Malley, Malley, and Pajevic 2011, 43). As Malley et al. suggest, "it can be applied after a more complex learning machine has done the heavy lifting of identifying an important set of predictors given a very large list of candidate predictors" (43). Especially in comparison with more complicated models, logistic regression models are relatively easy to interpret because they

Table 4.2
Sample of highly cited scientific publications referring to "logistic regression" in title or keyword

Year	Title	Keywords	Citations
2001	Prognostic Modeling With Logistic Regression Analysis: In Search Of A Sensible Strategy In Small Data Sets	Regression Analysis; Logistic Models; Bias; Variable Selection; Prediction	208
2002	Logistic Regression And Artificial Neural Network Classification Models: A Methodology Review	Artificial Neural Networks; Logistic Regression; Classification; Model Comparison; Model Evaluation; Medical Data Analysis	126
2002	A Comparison Of Statistical Approaches For Modelling Fish Species Distributions	Artificial Neural Networks; Classification Trees; Discriminant Analysis; Logistic Regression; Species Presence/absence	132
2002	New Strategies For Identifying Gene Gene Interactions In Hypertension	Data Reduction; Epistasis; Essential Hypertension; Gene Gene Interaction; Hardy Weinberg Disequilibrium; Linkage Disequilibrium; Logistic Regression; Multifactor Dimensionality Reduction; Pattern Recognition	200
2002	Mapping Epistemic Uncertainties And Vague Concepts In Predictions Of Species Distribution	Epistemic And Linguistic Uncertainty; Generalized Linear Models; Logistic Regression; Vagueness; Confidence Intervals; Prediction; Visualization; Model	106
2003	A Simple And Efficient Algorithm For Gene Selection Using Sparse Logistic Regression	NANA	111
2004	Validation And Updating Of Predictive Logistic Regression Models: A Study On Sample Size And Shrinkage	Logistic Regression; Validation; Updating; Shrinkage	156
2004	Classification Of Gene Microarrays By Penalized Logistic Regression	Cancer Diagnosis; Feature Selection; Logistic Regression; Microarray; Support Vector Machines	109
2005	Logistic Regression Model To Distinguish Between The Benign And Malignant Adnexal Mass Before Surgery: A Multicenter Study By The International Ovarian Tumor Analysis Group	NANA	125
2005	Logistic Model Trees	Model Trees; Logistic Regression; Classification	161
2005	Sparse Multinomial Logistic Regression: Fast Algorithms And Generalization Bounds	Supervised Learning; Classification; Sparsity; Bayesian Inference; Multinomial Logistic Regression; Bound Optimization; Expectation Maximization (em); Learning Theory; Generalization Bounds	176

(continued)

Table 4.2
Sample of highly cited scientific publications referring to "logistic regression" in title or keyword (*continued*)

Year	Title	Keywords	Citations
2005	Fauna Habitat Modelling And Mapping: A Review And Case Study In The Lower Hunter Central Coast Region Of Nsw	Bootstrapping; Conservation Planning; Habitat Modelling; Logistic Regression; Model Evaluation; Roc	128
2006	Modelling Distribution And Abundance With Presence Only Data	Case Control; Distribution; Habitats; Logistic Discrimination; Logistic Regression; Presence Only Studies; Pseudo Absences; Resource Selection Functions; Rsf; Sampling	191
2007	Random Forests For Classification In Ecology	Additive Logistic Regression; Classification Trees; Lda; Logistic Regression; Machine Learning; Partial Dependence Plots; Random Forests; Species Distribution Models	563
2007	Large Scale Bayesian Logistic Regression For Text Categorization	Information Retrieval; Lasso; Penalization; Ridge Regression; Support Vector Classifier; Variable Selection	147
2007	An Interior Point Method For Large Scale L(1) Regularized Logistic Regression	Logistic Regression; Feature Selection; L(1) Regularization; Regularization Path; Interiorpoint Methods	159
2008	Liblinear: A Library For Large Linear Classification	Large Scale Linear Classification; Logistic Regression; Support Vector Machines; Open Source; Machine Learning	316
2009	Genome Wide Association Analysis By Lasso Penalized Logistic Regression	NANA	191

superimpose the logistic function on the linear model that we have discussed already (see chapters 2 and 3). As *Elements of Statistical Learning* puts it, "the logistic regression model arises from the desire to model the posterior probabilities of the K classes via linear functions in x, while at the same time ensuring that they sum to one and remain in $[0, 1]$" (Hastie, Tibshirani, and Friedman 2009, 119). The logistic regression model predicts what class or category a particular instance is likely to belong to but "via linear functions in x."

We see something of this predictive desire from the basic mathematical expression for logistic regression in a situation where there are binary classes or $K = 2$:

$$Pr(G = K|X = x) = \frac{1}{1 + e^{\sum_{l=1}^{K-1}(\beta_{l0} + \beta_l^T x)}} \tag{4.2}$$

(Hastie, Tibshirani, and Friedman 2009, 119)

Equation 4.2 encapsulates lines in curves. That is, the linear model (the model that fits a plane to a scattering of points in vector space) appears as $\beta_l 0 + \beta_l^T x$, where as usual β refers to the parameters of the model and x to the matrix of input values. The linear model has, however, now been relayed through the sigmoid function so that its output values no longer increase and decrease linearly. Instead, they follow the curve of the logistic function and range between a minimum of 0 and a maximum of 1, a range of values that map onto probabilities (as discussed in chapter 5). Small typographic conventions diagram some of this transformation. In equation 4.2, new characters appear: G and K. Previously, the response variable, the variable the model is trying to predict, appeared as Y. Y refers to a continuous value, whereas G refers to membership of a group or class (e.g., survival vs. death, male vs. female, etc.).[9]

9 What does this wrapping of the linear model in the curve of the sigmoid logistic curve do in terms of finding a function? Note that the shape of this curve has no intrinsic connection or origin in the data. The curve no longer corresponds to growth or change in size, as it did in its 19th-century biopolitical application to the growth of populations. Rather, the curvilinear encapsulation of the linear model allows the left-hand side of the expression to move into a different register. The left-hand side of the expression is now a probability function and defines the probability (Pr) that a given response value (G) belongs to one of the predefined classes ($k = 1, \ldots, K-1$). In this case, there are two classes ('yes/no'), so $K = 2$. Unlike linear models, which predict continuous y values for a given set of x inputs, the logistic regression model produces a probability that the instance represented by a given set of x values belongs to a particular class. When logistic regression is used for classification, values greater than 0.5 are usually read as class predictions of

Curves and the variation in models

Whether the logistic function is a useful approximation to "the function that underlies the predictive relationship between input and out" depends on how it relates input and output. The way in which we have learned the logistic function by taking a textbook formula expression of it and plotting the function associated with it is not the way that machine learners typically learns an approximation to the predictive relationship between the input data and the output or response variable. For a machine learner, finding a function means optimizing function parameters on the basis of the data not deriving a formula. Machine learning is not a matter of mathematical analysis but of algorithmic optimization.[10]

"yes," "true," or `1`. As a result, drawing lines through the vector space can effectively become a way of classifying things. The increase in flexibility comes at the cost of a loss of direct connection between the data or features in the generalized vector space, and the output, response, or predicted variables. They are now connected by a mapping that passes through the somewhat more mobile and dynamic operation of exponentiation *exp*, a function whose rapid changes can be mapped onto classes and categories.

10 Even in machine learning, some function-finding through solving systems of equations occurs. For instance, the closed form or analytical solution of the least sum of squares problem for linear regression is given by $\hat{\beta} = (\mathbf{X}^T\mathbf{X})^{-1}\mathbf{X}^T\mathbf{y}$. As we saw in the previous chapter, this expression provides a quick way to calculate the parameters of a linear model given a matrix of input and output values. This formula is derived by solving a set of equations for the values $\hat{\beta}$, the estimated parameters of the model. But how do we know whether a model is a good one, or that the function that a model proffers to us fits the functions in our data, or that it "minimizes the probability of error"? One problem with closed-form or analytical solutions typical of mathematical problem solving is that their closed form obscures the algorithmic processes needed to actually compute results. The closed-form solution estimates the parameters of the linear model by carrying out a series of operations on matrices of the data. These operations include matrix transpose, several matrix multiplications (so-called inner product), and matrix inversion (the process of finding a matrix that when multiplied by the input matrix yields the identity matrix, a matrix with 1 along the diagonal, and 0 for all other values). All of these operations take place in the vector space. When the dataset, however, has a hundred or a thousand rows, these operations can be implemented and executed easily. But as soon as datasets become much larger, it is not easy to actually carry out these matrix operations, particularly the matrix inversion, even on fast computers. For instance, a dataset with a million rows and several dozen columns is hardly unusual today. Although linear algebra libraries are carefully crafted and tested for speed and efficiency, closed-form solutions, even for the simplest possible structures in the data, begin to break down.

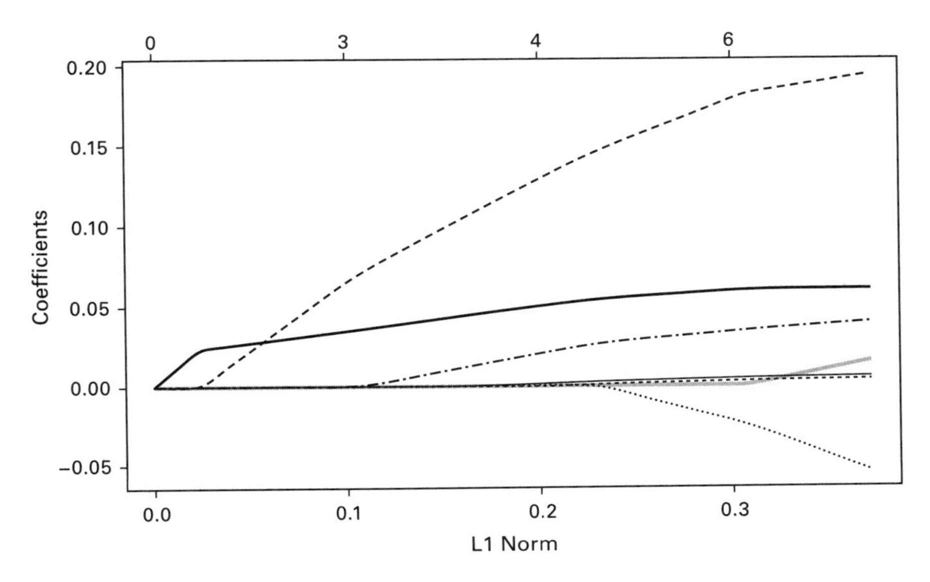

Figure 4.3
South African Heart Disease regularization plot.

If we turn just to the diagrammatic forms associated with logistic regression in *Elements of Statistical Learning,* something quite different and much more complicated than calculating the values of a known function presents itself there. For instance, in their analysis of the `South African coronary heart` disease data, Hastie and co-authors repeatedly model the risk of occurrence of heart disease using logistic regression. They first apply logistic regression fitted by "maximum likelihood," and then by "L1 regularized logistic regression" (Hastie, Tibshirani, and Friedman 2009, 126). The results of this function-finding work appear as tables of coefficients or "regularization plots." As is often the case, *Elements of Statistical Learning* assumes that readers already understand conventional statistical usages of logistic regression. Discussion dwells instead on observing how values of the model parameter change as different variants of the model transform the data.

Figure 4.3 shows a series of lines. Plotted after the model transforms the data 366 times, each line sets out the changing importance of a particular variable—`obesity, alcohol consumption, weight, age,` designated by numbers shown on the right-hand side—as it is included in the logistic regression model in a different way. The learning or function-finding diagrammed in figures 4.3 and 4.4 concerns variations in parameters and ways of automating the variation of parameters beyond

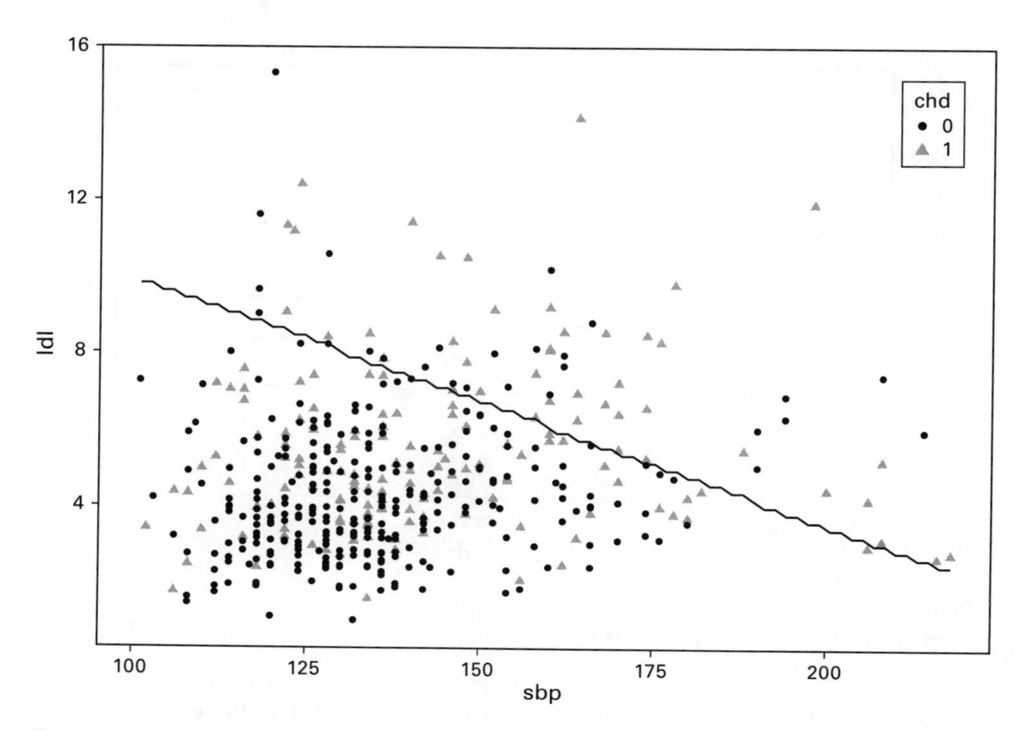

Figure 4.4
South African Heart Disease decision plane.

that undertaken by modeling experts such as statisticians and scientists when they fit models to data.[11]

We saw that the classic statistical model of linear regression fits lines to the data through the linear algebra method of ordinary least squares (see chapter 3,

11 As we saw in chapter 3, the production of new tables of numbers that list the parameters of models actually transform the vectorized data into new subspaces (a line, plane, surface). Many of the plots and tables found in machine learning texts, practice, and code offer nothing else but measurements of how the model parameters weight slightly different transformations of the vector space. In the case of logistic regression, the shape of the curve is determined using maximum likelihood. For present purposes, the statistical significance of this procedure is less important than the algorithmic implementation. This is the opposite of what might appear in a typical statistics textbook where the implementation of maximum likelihood would normally be quickly passed over. For instance, in *An Introduction to Statistical Learning with R*, a textbook focused on using R to implement machine learning techniques, the authors write: "we do not need to concern ourselves with the details of the maximum likelihood fitting procedure" (James et al. 2013, 133).

equation 3.3). Several obstacles hinder the construction of models using closed-form approximate solutions. Although closed form or analytical solutions do exist for linear regression, they don't exist for logistic regression or for more complex machine learners. Equally problematic, the closed-form solution is run once, and the model it produces is subject to no further variation. The parameters define the line of best fit. It can be interpreted by the modeler in terms of p or R^2 or other measures of the model's fit, but the model does not generate variations.[12]

Observing costs, losses, and objectives through optimization

Faced with the impracticality of an analytical or mathematically closed form solution to the problem of finding a function, machine learners typically seek ways of observing how different models traverse the data. They replace the exactitude and precision of mathematically-deduced closed-form solutions with algorithms that generate varying solutions. A range of techniques search for optimal combinations of parameters. These optimization techniques are the operational underpinning of machine learning. Without their iterative processes, there is no machine in machine learning. They have names such as "batch gradient descent," "stochastic gradient ascent," "coordinate descent," "coordinate ascent," and "Newtown-Raphson method" or simply "convex optimization" (Boyd and Vandenberghe 2004). These techniques have a variety of provenances (Newton's work in the 17th century, for instance, but more typically fields

12 As soon as we move from the more theoretical or expository accounts of function-finding into the domain of practice, instruction, and learning of machine learning, a second sense of function comes to the fore. The second sense of function comes from programming and computer science. A function there is a part of the code of a program that performs some operation, "a self-contained unit of code," as Derek Robinson puts it (Robinson 2008, 101). The three lines of R code written to produce the plot of the logistic function are almost too trivial to implement as a function in this sense, but they show something of the transformations that occur when mathematical functions are operationalised in algorithmic form. The function is wrapped in a set of references. First, the domain of x values is made much more specific. The formulaic expression $f(x) = 1/(1+e^{-x})$ says nothing explicitly about the x values. They are implicitly real numbers (i.e., $x \in \mathbb{R}$) in this formula but in the algorithmic expression of the function they become a sequence of 20001 generated by the code. Second, the function is flattened into a single line of characters in code, whereas the typographically the mathematical formula had spanned two to three lines. Third, a key component of the function $e^{-}x$ refers to Euler's number e, which is perhaps the number most widely used in contemporary sciences due to its connection to patterns of growth and decay (as in the exponential function e^x where $e = 2.718282$ approximately). This number, because it is "irrational," has to be computed approximately in any algorithmic implementation.

such as operations research that were the focus of intense research efforts during and after World War; see Bellman 1961; Petrova and Solov'ev 1997; Meza 2010). Much of the learning in machine learning occurs through these somewhat low-profile yet computationally intensive techniques of optimization.

Optimization is a practice of observation. "Science brings to light partial observers in relation to functions within systems of reference," write Gilles Deleuze and Félix Guattari in their account of scientific functions (Deleuze and Guattari 1994, 129).[13] In many machine learning techniques, the search for an approximation to the function that generated the data is optimized by reference to another function called the "cost function" (also known as the "objective function" or the "loss function"; the terms are somewhat evocative of both economics and cybernetics). Machine learning problems are framed in terms of minimizing or maximizing the cost function. Cost or loss takes the form of errors, and minimizing the cost function implies minimizing the number of errors made by a machine learner. As we saw earlier, in his formulation of the learning problem, the learning theorist Vladimir Vapnik speaks of choosing a function that approximates to the data yet minimizes the "probability of error" (Vapnik 1999, 31).

The cost function compares predictions generated by a machine learner to known values in the dataset. Every cost function implies some measure of the difference or distance between the prediction and the values actually measured. Cost functions in common use include squared error, hinge loss, log-likelihood, and cross-entropy. In classifying outcomes into two classes (the patient survives vs. patient dies, the user clicks vs. user doesn't click, etc.), the cost function has to express their either/or outcome. Crucially, if cost functions reconfigure "the act of fitting a model to data as an optimization problem" (Conway and White 2012, 183), function finding and hence machine learning in general occur iteratively. Given a cost function, a machine learner can vary its parameters keeping in view—or partially observing—whether the cost function increases or decreases. Just as the logistic function wraps the linear regression model in a sigmoid curve that switches smoothly between binary values, the cost functions diagram model parameters (usually noted as β) in relation to known responses or output values in the data. If there is learning here, it derives from mathematic forms or higher abstraction. Cost functions diagram relations between models and render their predictive reference through the negative feedback loops described by Norbert Wiener

13 Its hard to know whether Deleuze and Guattari were aware of the extensive work done on problems of mathematical optimization during the 1950 and 1960s, but their strong interest in the differential calculus as a way of thinking about change, variation, and multiplicities somewhat unexpectedly makes their account of functions highly relevant to machine learning.

(Wiener 1961). Importantly, these feedback loops are not closed mechanisms but places from which variations can be viewed.

For instance, the log-likelihood function, a typical and widely used cost function associated with logistic regression is defined as:

$$J(\beta) = \sum_{i=1}^{m} y_i log h(x_i) + (1 - y_i) log(1 - h(x_i)) \tag{4.3}$$

where

$$h_\beta(x) = \frac{1}{1 + e^{-\beta^T x}}$$

Equation 4.3 enfolds several manipulations and conceptual framings (particularly the Principle of Maximum Likelihood, a statistical principle; see chapter 5). But key terms stand out. First, the cost function $J(\beta)$ is a function of all the parameters (β) of the model. The parameters enter through the subsidiary function $h_\beta(x)$, the logistic function function encapsulating a linear function $\beta^T X$. Second, the function defines a goal of maximizing the overall value of the expression $J(\beta)$ as a function of variations in the parameters β. Third, the heart of the cost function is balancing two tendencies: it adds ($\sum$) all the values where the probability of the predicted class of a particular case $h(x_i)$ matches the actual class y_i and subtracts $(1-y)$ all the cases where the probability of the predicted class does not match the actual class. This so-called *log likelihood* function can be maximized through optimization, but not solved in closed form. The optimal values for β, the model parameters that define the model function, need to be found through some kind of search.

Gradients as partial observers

We have some sense of how a function can be configured as an observer but little sense of how they manage variations. Many optimization techniques rely on differential calculus and particularly the calculus of variations to maximize or minimize the value of a cost function. In fact, loss functions are often chosen on the basis of their differentiability. Many machine learners use an optimization algorithm called "gradient descent." In neural nets and deep learning, gradient descent (or ascent) occurs on an increasingly vast scale. It optimizes the parameters of a model by searching for the maximum or minimum values of the objective function. The algorithm can be written using calculus-style notation as:

$$\text{Repeat until convergence: } \beta_j := \beta_j + \alpha(y_i - h_\beta(x_i))x_{\beta j} \tag{4.4}$$

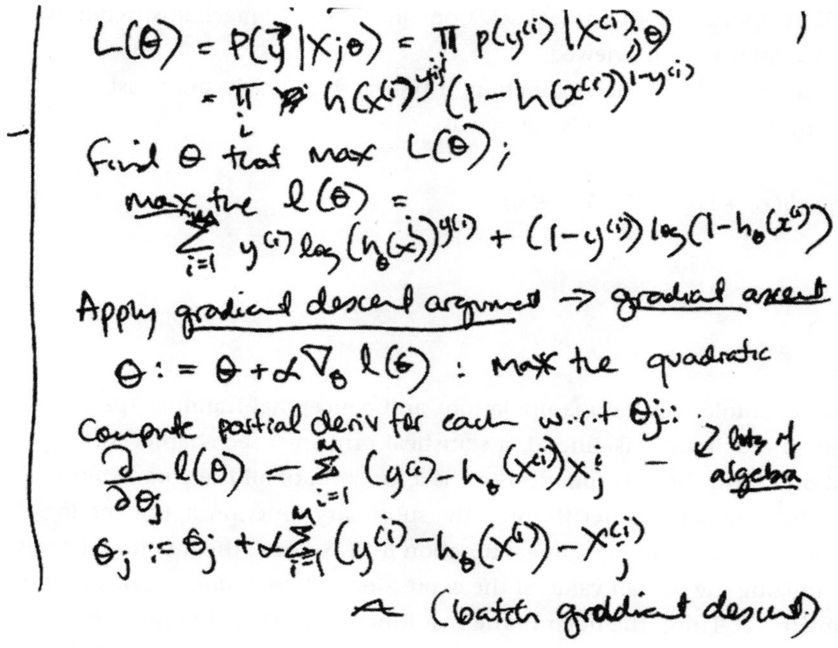

$L(\theta) = P(\vec{y}|X;\theta) = \prod p(y^{(i)}|x^{(i)};\theta)$

$= \prod h(x^{(i)})^{y^{(i)}}(1-h(x^{(i)}))^{1-y^{(i)}}$

Find θ that max $L(\theta)$;

max the $l(\theta) =$

$\sum_{i=1}^{m} y^{(i)} \log(h_\theta(x^{(i)}))^{y(i)} + (1-y^{(i)}) \log(1-h_\theta(x^{(i)}))$

Apply gradient descent argument → gradient ascent

$\theta := \theta + \alpha \nabla_\theta l(\theta)$: max the quadratic

Compute partial deriv for each w.r.t θ_j:

$\frac{\partial}{\partial \theta_j} l(\theta) = \sum_{i=1}^{m} (y^{(i)} - h_\theta(x^{(i)}))x_j^{(i)}$ — ← lots of algebra

$\theta_j := \theta_j + \alpha \sum_{i=1}^{m} (y^{(i)} - h_\theta(x^{(i)}) - x_j^{(i)}$

(batch gradient descent)

Figure 4.5
Gradient ascent for logistic regression.

The version of the algorithm shown in algorithm 4.4 is called "stochastic gradient descent." Archaeologically, in excavating such formula, the point is not to read and understand them directly but to characterize the enunciative function that regulates them. Practical understanding would be the point in a machine learning course. Actually reading these formal expressions and being able to follow the chain of references and indexical signs that lead away from them in various directions depend very much on the diagrammatic processes described in chapter 2. Many people who directly use machine learning techniques in industry and science would not often if ever need to make use of such expressions as they build models. They would mostly take them for granted and simply execute via functions supplied by software libraries (e.g., `GradientDescentOptimizer` in the `TensorFlow` library or `StochasticGradient` in `torch`).

Given that equation 4.4 encapsulates the heart of a major optimization technique, we might first of all be struck by its operational brevity. This algorithm is not elaborate or convoluted. As Malley, Mally, and Pajevic observe, "most of the [machine learning] procedures ... are (often) nearly trivial to implement" (Malley, Malley, and Pajevic

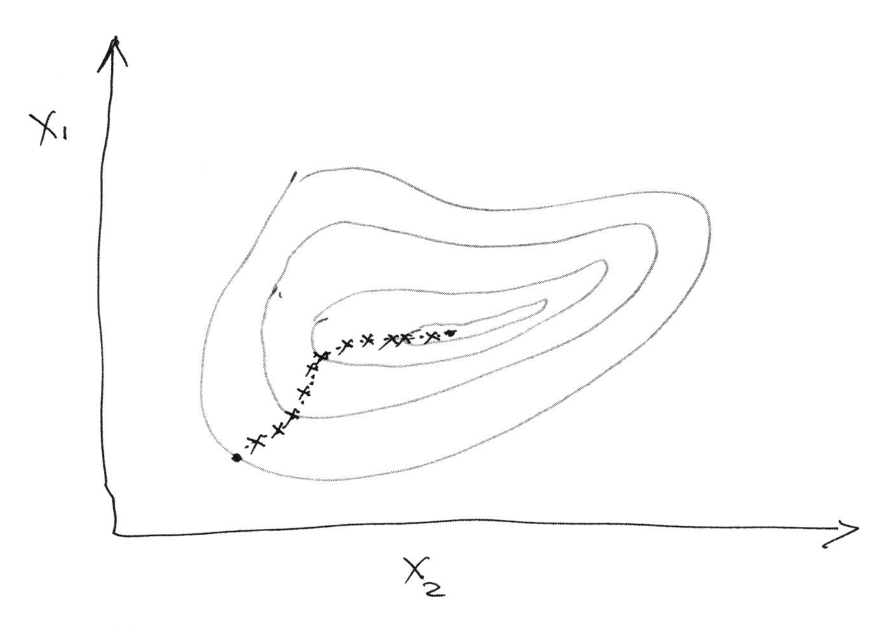

Figure 4.6
Stochastic gradient descent path.

2011, 6). Note that this expression of the algorithm, taken from the class notes for [Lecture 3] of Andrew Ng's "Machine Learning" CS229 course at Stanford (Ng 2008g; see figure 4.5), mixes an algorithmic set of operations with function notation. We see this in several respects: the formulation includes the imperative "repeat until convergence"; it also uses the so-called "assignment operator" $:=$ rather than the equality operator $=$. The latter specifies that two values or expressions are equal, whereas the former specifies that the values on the right-hand side of the expression should be assigned to the left. Both algorithmic forms—repeat until convergence and assign/update values—owe more to techniques of computation than to mathematical abstraction.

In gradient descent, we see functions acting as partial observers. The specification for the gradient descent algorithm brings us to the scene where ongoing transformation of data from irregular volume to plane can be observed. At the heart of this reshaping lies a different mathematical formalism: the partial derivative, $\frac{\partial}{\partial \beta_j} J(\beta_j)$. Like all derivatives in calculus, this expression can be interpreted as the rate at which one variable changes in relation to another; that is, as the rate at which the cost function $J(\beta)$ changes with respect to the different values of β.[14] Much learning in machine learning pivots on

14 The derivative $\frac{\partial}{\partial \beta_j} J(\beta_j)$ is *partial* because β is a vector $\beta_0, \beta_1 \cdots \beta_j$.

the observation of rates of change of a cost function in relation to its arguments, the j-dimensional vector space defined by β, the model parameters. The partial derivatives in the gradient descent algorithm observe the direction in which the value of the cost function reduces. Each iteration of the algorithm reduces or increases the parameters β of the model in the direction of reduced cost and perhaps less error. Importantly, the derivative of a sigmoid (or logistic function) $\sigma(x)$ is given by $\sigma(x)(1 - \sigma(x))$, which means that the partial derivatives of a cost function will be easy to compute.

The power to learn

The power of machine learning to learn, its power to epistemologize, pivots around functions in disparate yet connected ways: the transformation of data through operational functions maps new subspaces in vector space and observational functions algorithmically superimpose new constraints—cost, loss, or objective functions—that direct an iterative process of optimization. Machine learning diagrammatically distributes learning in the operational human–machine formation. People look at curves for evidence of convergence, functions compress data into functions that support classification or predictions, and algorithms observe gradients or rates of error in relation to model parameters. In several senses, people and machines together move along curves. The logistic function folds the lines that best fit the data into a probability distribution that can be read in terms of classification. The cost functions, as they seek to minimize differences between the predicted values and the known values found in the vector space, control variations in the model. Every observer in this domain is partial because the humans cannot see lines or curves in the multidimensional data, the functions that underpin models such as logistic regression or linear regression can transform data in the vector space, but can't show how well they see it, and the processes of optimization only see the results of the model and its errors, not anything in its referential functioning. Omniscience (*apropos* master algorithms), whether fully supervised or completely unsupervised, is impossible here.

Amid this endemic partiality, we can begin to understand the multiplication of machine learners and the mirage of universality. Machine learners are functions that transform data and observe the effects of those transformation in learning to classify, predict, and rank. But the function that defines a machine learner contracts a range of partial observers relaying values and changes to each other. The operational power of machine learning depends on the diagrammatic and sometimes experimental relays between different practices of observing. Attending to specific mathematical functions in isolation—the logistic function, the Lagrangean, the Gaussian, the quadratic discriminant, and so on—will not tell us how the operational power of functions comes together in machine learning, but it may provide ways of mapping the diagrammatic connections, the enunciative function that connects different elements in the production of consequential classifications and predictions, generating operational statements in fields of knowledge.

We are in a slightly better position to understand now how there can be many machine learners but a relative sparsity in the production of statements. Gradients—continuous variations in rate—are useful because they generate many functions, many approximations within one operational process, within one enunciative function. The profusion of machine learners, the "bewildering variety" that Domingos and others celebrate and flag up, can be seen as the effect of an operational formation predicated on approximation through variation.

In his account of Foucault's diagrams of power, Gilles Deleuze writes:

> every diagram is intersocial and constantly evolving. It never functions in order to represent a persisting world but produces a new kind of reality, a new model of truth.... It makes history by unmaking preceding realities and significations, constituting hundreds of points of emergence or creativity, unexpected conjunctions or improbable continuums (Deleuze 1988b, 35).

Functions in machine learning are "intersocial" in the sense that they bring together different mathematical, algorithmic, operational, and observational processes. The sigmoid function switches the geometry of the linear model over into the calculation of probabilities and classification but also eases the calculation of partial derivatives. Cost functions recraft statistical modeling as a quasi-iterative process of model generation and comparison. New kinds of realities arise in which the classifications and predictions generated by the diagonal connections between mathematical functions and operational processes of optimization can constitute a "new model truth" and can unmake "preceding realities and significations." Despite my deliberately narrow focus on a single set of relays that connect linear models, the logistic function, the cost function, and gradient ascent, hundreds and perhaps hundreds of thousands of "points of emergence" associated with this diagram of functioning.

The machine learning diagram, like any functioning, harbours the potential for invention. Describing the application of machine learning to biomedical and clinical research, James Malley, Karen Malley, and Sinisa Pajevic contrast it to more conventional statistical knowledges:

> working with statistical learning machines can push us to think about novel structures and functions in our data. This awareness is often counterintuitive, and familiar methods such as simple correlations, or slightly more evolved partial correlations, are often not sufficient to pin down these deeper connections (Malley, Malley, and Pajevic 2011, 5–6).

Novel structures and functions in "our data" are precisely the functions that machine learning technique seek to learn. Could new habits or actively diverging worlds that Stengers calls for appear amid this quasi-iterative pursuit of optimization and convergence? This is a terrain for critical thought to explore. A function in isolation never learns. But when watched or observed, even virtually, divergence has some chance. To the extent that machine learners relay references experimentally between things and people, mobilizing the production of statements and visibilities across different elements, divergence remains possible.

5 $N = \forall X$: Probabilization and the Taming of Machines

In the final pages of *The Taming of Chance*, Ian Hacking sums up the work of the philosopher Charles Sanders Peirce as a two-way affirmation of chance. First, Peirce, following the work of the psychophysicist Gustav Fechner and before him the astronomer-sociologist Adolphe Quetelet, ontologically reconfigured the normal curve.[1] The "personal equation," the variation in measurements made by any observer, becomes "a reality underneath the phenomena of consciousness" (Hacking 1990, 205). Peirce's belief in absolute chance or a stochastic ontology, "a universe of chance" as Hacking puts it, continued a series of realizations of curves, in which astronomical, social, biological, and finally psychological variations were all understood as generated by processes of chance. Second, and to show the underlying reality of the normal curve, "Peirce deliberately used the properties of chance devices to introduce a new level of control into his experimentation. Control not by getting rid of chance fluctuations, but by adding some more" (Hacking 1990, 205). In the century or so since, what happened to the thorough-going affirmation of statistical thought and probabilistic practice epitomized by Peirce? Hacking stresses that he does not understand Peirce as the precursor or innovator of 20th-century statistical thought (Hacking's *Taming of Chance* ends at 1900) but rather as "the first philosopher to conceptually internalize the way chance had been tamed in the nineteenth century" (215). What would the equivalent philosopher-machine learner internalize today? What would such persons, working in science or media or government, hold firm in relation to chance, probability, and statistics?

In the opening lines of the preface to the First Edition of *Elements of Statistical Learning*, Hastie, Tibshirani, and Friedman describe the altered situation of statistics:

1 The historian of statistics Stephen Stigler provides a lengthy account of Fechner's work in (Stigler 1986, 239–259).

The field of Statistics is constantly challenged by the problems that science and industry brings to its door. In the early days, these problems often came from agricultural and industrial experiments and were relatively small in scope (Hastie, Tibshirani, and Friedman 2009, xi).

(At the end of the preface, they also cite, we might note in passing, Hacking's work: "The quiet statisticians have changed our world" (Hastie, Tibshirani, and Friedman 2009, xii).) One of the challenges that science and industry has brought to the door of statistics in recent years has not only been more data but also machine learners. What difference do the "vast amounts of data ... generated in many fields" (xi) make to the field of statistics? Statistics has, I will suggest in this chapter, gradually *probabilized* machine learners or injected a substratum of chance that flows directly from their operation. To grasp this *probabilization*, we need to determine what role randomness, change, and the probabilistic distribution of elements and events play in machine learning. These questions of how worlds become thinkable through machine learning can be addressed partly by contrasting the "taming of chance" achieved by 18th- and 19th-century statistics and the taming of data—and machines—in statistical practices of machine learning today.

Data reduce uncertainty?

The broadest claim associated with machine learning hinges on the simple expression shown below:

$$N = \forall X \tag{5.1}$$

In equation 5.1, N refers to the number of observations (and hence the size of the dataset); the logical operator $\forall$ means "all" because this is the level of inclusion that many fields of knowledge in science, government, media, commerce, and industry envisage; and X refers to the data arrayed in vector space. Note that this expression leaves some things out. Y, the response variable, for instance, may or may not be known or part of the data X. Although both the expansion of data in the vector space and the machine learners who transform and observe it have appeared in previous chapters, I focus here on changes in probability practices associated with machine learning, and in particular $N = \forall X$, the claim that with all the data, the production of knowledge fundamentally changes.

The claim that with $N = \forall X$ the nature of knowledge changes has been widely discussed.[2] Viktor Mayer-Schönberger and Kenneth Cukier's *Big Data: A Revolution*

2 Rob Kitchin provides a useful overview of these claims (Kitchin 2014). Although I will not analyze the claims about big data in specific cases in any great detail, the growing literature on

That Will Transform How We Live, Work and Think presents this shift in many different settings in the course of the vignettes and teeming comparisons that have become typical of the data revolution genre. In a chapter entitled "More," they sketch the transition from data practices reliant on sampling to data practices that deal with all the data:

> Using all the data makes it possible to spot connections and details that are otherwise cloaked in the vastness of the information. For instance, the detection of credit card fraud works by looking for anomalies, and the best way to find them is to crunch all the data rather than a sample (Mayer-Schönberger and Cukier 2013, 27).

In the several hundred pages that follow in *Big Data*, the problem of how to "crunch all the data" is not a major topic. Machine learning remains almost completely invisible as a practice of transforming data in the name of knowledge. Although they mention the role of social network theory (30), "sophisticated computational analysis" (55), "predictive analytics," (58) and "correlations" (7), and they observe that "the revolution" is "about applying math to huge quantities of data in order to infer probabilities" (12), any further consideration of a change in data practices is largely confined to a business-oriented contrast between having some of the data and having all the data (i.e., businesses often have all the data on their customers).

Without a sense of how statistical practices animate and configure key features of "crunching the data" to make predictions, it becomes hard to see how the "revolution" takes place. Just as 19th-century statistics transformed many measurements into population attributes (e.g., mean as the ideal or abstract property of a population), the shift between n and $\forall \boldsymbol{X}$, between some and all, a shift dependent on machine learning, internalizes, I will suggest, population attributes into the operations of machine learners. This statistical event is akin to the advent of the Normal distribution (and indeed N is a standard symbol for the Normal distribution in statistics textbooks) as a way of knowing and controlling populations (Hacking 1975, 108). To signal its continuity with the invention of probability, I term it "probabilization," a pleonasm that refers to facets of the operational formation that address and configure machine learners in terms of probabilities.

---

this topic suggests that machine learning in its various operations—epistopic construction of vector space, function finding as association of partial observers, and a reinternalization of probability—generates considerable difficulties and challenges for knowledge, power, and production.

Machine learning as statistics inside out

The argument mimics Hacking's. In *The Taming of Chance*, Hacking argues that modern statistical thought transposed a way of calculating errors in experimental measurements and astronomical observations into the real and essential attributes of populations understood as processes of reproductive growth. This transposition or inversion relied on four intermediate steps passing through the development of a probability calculus (particularly the work of Jacob Bernoulli and the binomial or heads-tails probability distribution in the 1690s (Hacking 1975, 143)), the accumulation of large numbers of measurements (the most famous being the chest measurements of soldiers in Scottish regiments, but these were only one flurry amid an avalanche of numbers in the 1830 and 1840s), the emergence of the idea of multiple, minute, independent causes producing events (particularly as developed in medicine but also in studies of crime), and the "law of errors" applying to measurements made by, among others, astronomers (Hacking 1990, 111–112). As Hacking observes, coins, suicides, crime, chest measurements, and astronomical observations all pile up in a statistical aggregate that remains, although somewhat altered, indelible in contemporary statistical knowledges, particularly in its frequent recourse to notions of population, probability, and distribution. In this aggregate, observers and the observed changed places. The distribution of errors made by astronomers measuring the position of stars or planets became a distribution or variation inherent in a population.

Machine learning reverse-engineers the invention of modern statistical thinking. It takes back the "real quantities"—probabilities—that modern statistics had attributed to the populations in the world and distributes them to devices, to machine learners that people then observe, monitor, and indeed measure again in many ways. The direct swapping between uncertainty in measurement and variation in real attributes that statistics achieved now finds itself rerouted and intensified as machine learners measure the errors, the bias, and the variance of devices. Although it relies heavily on probability distributions, machine learning is a fat-tailed distribution of probability.

The swapping or redistribution is not a simple mirror-image reversal, as if machine learners mistake devices for a population. Machine learning frequently premises statistical thinking as a basic condition for its operations and devices. When *Elements of Statistical Learning* states that (as we saw in the previous chapter) "our goal is to find a useful approximation $(\hat{f})(x)$ to the function $f(x)$ that underlies the predictive relationship between input and output" (Hastie, Tibshirani, and Friedman 2009, 28), they invoke the "real quantities" first elaborated and articulated by proto-statisticians, such

Table 5.1
Some structuring differences in machine learning

parametric	non-parametric
bias	variance
prediction	inference
generative	discriminative

as Quetelet grappling with population and sample parameters. The major structuring operational practices in machine learning show the marks of a strong commitment to the reality of the statistical and to the ongoing probabilization of machine learners.

What is probabilization in practice? Reading and working with machine learning techniques usually means encountering and responding to apparatus drawn from statistics, but the apparatus is not typically the statistical tests of significance or variation. In contrast to a statistics textbook such as the widely used *Basic Practice of Statistics* (Moore 2009) or a more advanced guide such as *All of Statistics* (Wasserman 2003), where statistical tests (t-test, chi-squared test, etc.), hypothesis testing, and analysis of uncertainties (confidence intervals, etc.) order the exposition, machine learning textbooks rely on a conceptual apparatus curiously stripped of statistical tests and measurements. Statistical underpinnings may be fundamental, but this does not mean that machine learners simply automate statistics.

Instead, a basic set of contrasts or indeed oppositions that owe much to probabilistic thinking order, compose, associate, and link the statements of machine learners. The contrasts shown in table 5.1 all have a statistical facet and anchoring to them. Some refer to errors that affect how a machine learner refers to data (bias and variance; see discussion below); some designate an underlying statistical intuition about how particular machine learners treat data (does the model seek to generate the data or classify—discriminate—it e.g., Naive Bayes or Latent Dirichlet Allocation are generative models, whereas logistic regression or support vector machines are *discriminative*); parametric and nonparametric describe the role of probability distributions in the model; and others indicate different kinds of statistical knowledge practice (prediction seeks to anticipate, whereas inference seeks to interpret, etc.; also see discussion below). These broad structuring differences reach down deeply into the architecture, diagrams, practices, statements, visual objects, and computer code associated with $N = \forall X$. Because they anchor basic operations of machine learning in probability, in the last two decades, formalisms derived from statistics have increasingly populated the field, furnishing and rearranging its diagrammatic references to the worlds of industry, agriculture, earth science, genomics, and so on, but also, crucially, triggering ontological mutations in machine learners themselves.

Distributed probabilities

Although these structuring differences deeply shape practice in machine learning, the underlying operator that allows swapping between knowledge and the world, between events and devices, is probability, and, in particular, functions that describe variations in populations, probability distributions. Probability distributions both map population variations and, as we will see, multiply the number of things that count as populations.

The normal distribution pervaded 19th-century statistical thinking as it targeted populations across law, medicine, agriculture, finance, and, not least, social life. Normal distributions appear in countless variations in scientific, government, and institutional settings as functions that collect events, measurements, observations, and records into evidential probability quantities.[3]

$$f(x;\mu,\sigma^2) = \frac{1}{\sigma\sqrt{2\pi}}e^{-\frac{1}{2}\left(\frac{x-\mu}{\sigma}\right)^2} \tag{5.2}$$

The function shown in equation 5.2 expresses the probability of a given value of the variable x given a population whose variations (with respect to x) can be expressed in terms of two parameters, μ and σ, the mean and variance. This is the so-called normal or Gaussian distribution.[4] Its mathematics were intensively worked over during the late 18th and early 19th centuries in what has been termed "one of the major success stories in the history of science" (Stigler 1986, 158). The normal distribution has a power-laden biopolitical history closely tied with knowledges and governing of

3 Statistical graphics have a rich history and semiology that I do not discuss here (see Bertin 1983).

4 Dozens of differently shaped probability distributions map continuous and discrete variations to real numbers. Other probability distributions—normal (Gaussian), uniform, Cauchy exponential, gamma, beta, hypergeometric, binomial, Poisson, chi-squared, Dirichlet, Boltzmann-Gibbs distributions, etc. (see NIST 2012 for a gallery of distributions)—functionally express widely differing patterns. The queuing times at airport check-ins do not, for instance, easily fit a normal distribution. Statisticians model queues using a Poisson distribution, in which, unfortunately for travellers, distributes the number of events in a given time interval quite broadly. Similarly, it might be better to think of the probability of rain today in northwest England in terms of a Poisson distribution that models clouds in the Atlantic queuing to rain on the northwest coast of England. (Rather than addressing the question of whether it will rain, a Poisson-based model might address the question of how many times it will rain today.)

populations in terms of morality, mortality, health, and wealth (see Hacking 1975, 113–124). The key parameters here include μ, the mean, and σ, the variance, a number that describes the dispersion of values of the variable, x. These two parameters together define the shape of the curve. Given knowledge of μ and σ, the normal or Gaussian probability distribution maps all outcomes to probabilities (or numbers in the range 0 to 1). Put statistically, functions such as the Gaussian distribution probabilize events as random variables. Every variable potentially becomes a function: "a random variable is a mapping that assigns a real number to each outcome" (Wasserman 2003, 19).

The possibility of treating population variations as random variables, that is, as probability distributions, was a significant historical achievement, one that continues to develop and ramify.[5] Random variables distribute probability in the world. When conceptualized as real quantities in the world rather than epiphenomenal byproducts of errors in our observations or those of measuring devices, probability distributions weave directly into the productive operations of power. Distribution in the sense of locating, positioning, partitioning, sectioning, serializing, or queuing operations has received much more attention in critical thought (particularly in the many uses of Foucault's concept of disciplinary power; Foucault 1977), but in almost every setting, distribution in the sense of counting, apportioning, and weighting of different outcomes also operates. This constant interweaving of spatial, architectural, logistical, and functional processes has energized statistical thought for several centuries.[6] For instance, given the normal distribution, it is possible, under certain circumstances, to effectively subjectify individuals. If an educational psychologist indicates to someone that his or her intelligence lies toward the left-hand side of the normal curve peak (and hence less than the population mean), then they quickly assign them to a potentially

5 The mapping that assigns numbers to outcomes (heads v. tails; cancer v. benign; spam v. not-spam) is a probability distribution. As I have argued in (Mackenzie 2016), random variables have become much more widespread in statistical practice due to changes in computational techniques.

6 "Distribution" pervades Foucault's account of power and knowledge from *The Order of Things* (Foucault 1992 [1966]) onward. Foucault treats distributions in several different ways: as spatial or logistical techniques, as mathematical orderings of large numbers of people or things, and as a methodological and theoretical framing device. In *Discipline and Punish* (Foucault 1977), the spatial sense prevails, but in later works, the population or demographic sense of distribution takes precedence (Foucault 1998). Distribution certainly has theoretical primacy in his account of power: "relations of power-knowledge are not static forms of distribution, they are 'matrices of transformations' " (Foucault 1998, 99).

institutionally and economically consequential trajectory. Since its inception in the social physics of Adolphe Quetelet as a defining property of populations, the normal curve has not only described but modulated and reshaped populations (in terms of health, morality, and wealth).

Given that functions such as equation 5.2 have persisted for so long as elements of population governmentality or biopolitics, what happens to them in machine learning? The pages of a book such as *Elements of Statistical Learning* show many signs of an ongoing invocation of probability distributions. We could simply observe their abundance. Hastie and co-authors frequently invoke probability distributions. They speak of "Gaussian mixtures," "bivariate Gaussian distributions," "standard Gaussian," "Gaussian kernels," "Gaussian assumptions," "Gaussian errors," "Gaussian noise," "Gaussian radial basis function," "Gaussian variables," "Gaussian densities," "Gaussian process," and so forth. (The term "normal" appears in an even wider spectrum of similar guises.) Events, things, properties, operations, functions, and attributes all associate with probability distributions.

The multiple invocations of probability distributions map a variety of events (occurrence of cancer, occurrence of the word "Viagra" in an email, a click on a hyperlink, etc.) to real numbers. Despite the sometimes dense mathematical diagrammaticism, the term *distribution* emphasizes a tangible and practically resonant way of thinking about how events or possible outcomes shift as the parameters of a function vary.[7] Whatever inferences and predictions become possible, probability distributions are a crucial control surface for machine learning understood as a form of movement through data. In contrast to the endowment of living aggregates, such as populations with probability that we see in the biopolitical history of statistics (and later in the natural sciences such as physics and biology), statistical machine learning increasingly constitutes devices as populations via probability distributions.

7 Machine learners adjust these parameters in different ways. For instance, parametric and nonparametric models (see table 5.1) differ in that the former have a limited number of parameters and the latter an undefined number of parameters (e.g., Naive Bayes, k nearest neighbours or support vector machine models). But both kinds assume that an underlying probability distribution—a function, "unobservable" or not—operates, even if it changes with new data. A probability distribution under these assumptions becomes the closest reality we have to whatever process generated all the variations in data gathered through experiments and observations. From a probabilistic perspective, the task of machine learning is to estimate the parameters (the mean μ and variance σ in the case of Gaussian curve) that shape the curve of the probability distribution.

Naive Bayes and the distribution of probabilities

How could machine learners become a population? The mathematical expression for one of the most popular of all machine learning classifiers, the Naive Bayes classifier, stands out for its probabilistic simplicity and seeming lack of "moving parts":

$$f_j(X) = \prod_{k=1}^{p} f_{jk}(X_k) \tag{5.3}$$

(Hastie, Tibshirani, and Friedman 2009, 211)

Some machine learners are so simple that they can be implemented in a few lines of code. Along with the perceptron, linear regression, and k nearest neighbors, the function shown in equation 5.3 is one of the simplest to be found in most machine textbooks yet easily adapts for high-dimensional data, the kind of data associated with contemporary network infrastructures, scientific instruments, online communications, and $N = \forall \boldsymbol{X}$ in general.[8] Although the Naive Bayes classifier is one of the most popular machine learning algorithms, it is more than 50 years old (Hand and Yu 2001).

The key diagrammatic element of the classifier in the equation is $\prod$, an operator that multiplies all the values of the matrix of X values (from 1 to p) to generate a product. What product does the Naive Bayes classifier produce? The expression $f_j(X)$ refers to a probability density; that is, it describes the probability that a particular thing (a document, an image, an email message, a set of URLs, etc.) belongs to the class of things j. In constructing an estimate of the probability that a given message, image, or event is an instance of class j, p different features are taken into account. (The subscript k indexes the p dimensions of the vector space.) The subscripts $k = 1$ on the $\prod$ operator and k on the data X_k indicate that the Naive Bayes classifier makes use of a series of

8 The other contender for simplest machine learner would be the also popular k nearest neighbors. As Hastie et al. observe: "these classifiers are memory-based and require no model to be fit" (Hastie, Tibshirani, and Friedman 2009, 463). Like the Naive Bayes classifier, the equation for k nearest neighbors is simple:

$$\hat{Y}(x) = \frac{1}{k} \sum_{x_i \in N_k(x)} y_i \tag{5.4}$$

where $N_k(x)$ is the neighborhood of x defined by the k closest points x_i in the training sample (Hastie, Tibshirani, and Friedman 2009, 14).

In equation 5.4, a parameter appears: k, the number of neighbors. This contrasts greatly with the linear models discussed in chapters 3 and 4, where the number of parameters p usually equals the number of variables in the dataset or dimensions in the vector space.

features or variables in calculating the overall probability that a given thing or observation belongs to a specific class. Put in the language of probability calculus, the classifier produces a probability density $f_j(X)$ by calculating the *joint probability* of all the *conditional* probabilities of the features or predictor variables in X for the class j. As *Elements of Statistical Learning* rather tersely puts it, "each of the class densities are products of the marginal densities" (Hastie, Tibshirani, and Friedman 2009, 108).

The Naive Bayes classifier directly invokes probability (including its name, with its reference to the Bayes Theorem), yet there is little obvious connection to statistics in its modern form of tests of significance. As Drew Conway and John Myles-White write in *Machine Learning for Hackers*,

> At its core, [Naive Bayes] . . . is a 20th century application of the 18th century concept of *conditional probability*. A conditional probability is the likelihood of observing some thing given some other thing we already know about (Conway and White 2012, 77).

They point to the application of "conditional probability," a probability conditioned on the probability of something else. Conditional probability lies at the heart of many of the data transformation associated with prediction or pattern recognition because it links a class to the occurrence of combinations of variables or features. Naive Bayes links variables by simply multiplying probabilities.[9] As any of the many accounts of the technique will explain, the name comes from Bayes Theorem, one of the most basic yet widely used results in probability theory (again dating from the 18th century), yet Naive Bayes does not even fully embrace Bayes Theorem as the principle of its operation. The classifier has a simple architecture based on the concepts of conditional probability and joint probability; it calculates a probability density function $f_j(X)$ or probability distribution for each possible class of things as a combination of the probabilities of all the many features or attributes of populations that come together in data. It makes the drastically simplifying assumption that features or variables are statistically independent of each other, where "independent" means that they do not affect each other, or that they have no relation to each other. We will see below that dramatic simplifications such as independence do not necessarily weaken the referential grasp of machine learners on the world but in certain ways allow them to reconfigure the operations of machine learners as a population of learners.

9 In Mackenzie (2014a), I have suggested that the intensification of multiplication associated with probabilistic calculation may constitute an important mutation in the ontological and practical texture of numbers. The epidemiological modelling of H1N1 influenza in London 2009 involved multiplying a great variety of probability distributions to calculate the conditional probability of influenza over time.

Spam: When $\forall N$ is too much?

$\forall N$, or having all the data, can be a bother. In *Doing Data Science*, Rachel Schutt and Cathy O'Neill furnish a `bash` script (i.e., command line instructions) to download the well-known `Enron` email dataset and build a Naive Bayes classifier that labels email as spam or not. In many ways, this is canonical machine learner pedagogy. For Naive Bayes, email spam detection has become the standard example (Andrew Ng uses it in CSS229, Lecture 5 (Ng 2008a)). In this setting, machine learners operate as filters coping with too much communication.

A typical spam email in the `Enron` dataset, a dataset that derives from the U.S Federal Energy Regulatory Commission's investigation into Enron Corporation (Klimt and Yang 2004), looks like this:

Subject: it's cheating, but it works ! can you guess how old she is ? the woman in this photograph looks like a happy teenager about to go to her high school prom, doesn't she ? she' s an international, professional model whose photographs have appeared in hundreds of ads and articles whenever a client needs a photo of an attractive, teenage girl.but guess what ? this model is not a teenager ! no, she is old enough to have a 7-year-old daughter.. she also says, "if it weren't for this amazing new cosmetic cream called 'deception,' i would lose hundreds of modeling assignments . . . because . . . there is no way i could pass myself off as a teenager." service dept 9420 reseda blvd # 133 northridge, ca 91324

The text of a typical non-spam email looks like this:

Subject: industrials suggestions. . . . . . ———————-forwarded by kenneth seaman / hou / ect on 01 / 04 / 2000 12 : 47 pm——————— - pat clynes @ enron 01 / 04 / 2000 12 : 46 pm to : kenneth seaman / hou / ect @ ect, robert e lloyd / hou / ect @ ect cc : subject : industrials ken and robert, the industrials should be completely transitioned to robert as of january 1, 2000.please let me know if this is not complete and what else is left to transition . thanks, pat

Such communications, with their mixture of solicitation and imperative, are familiar to anyone who uses email. How does Naive Bayes probablize their differences? How do they become X or even $f_j(X)$ in the Naive Bayes classifier? The code that *Doing Data Science* supplies is instructive.

The script draws out something of how the joint probability function in equation probabilizes a single word.[10] Not all machine learning models are so simple that they can be conveyed in 30 lines of code (including downloading the data and comments),

10 The input to the script is a single word such as "finance" or "deal." The model is so simple that it only classifies a single word as spam. The `bash` script carries out four different transformations of the data in building the model. It uses only command line tools such as `wc` (word count), `bc`

Listing 5.1

A Naive Bayes classifiers for `enron` email (Schutt and O'Neil, 2013, 105–106)

```
#!/bin/bash
# description: trains a simple one-word naive bayes spam
# filter using enron email data
# usage: ./enron_naive_bayes.sh <word>
# author: jake hofman (gmail: jhofman)

### PART 1
Nspam='ls -l spam/*.txt | wc -l'
Nham='ls -l ham/*.txt | wc -l'
Ntot=$Nspam+$Nham
echo $Nspam spam examples
echo $Nham ham examples

Nword_spam='grep -il $word spam/*.txt | wc -l'
Nword_ham='grep -il $word ham/*.txt | wc -l'
echo $Nword_spam "spam examples containing $word"
echo $Nword_ham "ham examples containing $word"

### PART 2
Pspam='echo "scale=4; $Nspam / ($Nspam+$Nham)" | bc'
Pham='echo "scale=4; 1-$Pspam" | bc'
echo
echo "estimated P(spam) =" $Pspam
echo "estimated P(ham) =" $Pham
Pword_spam='echo "scale=4; $Nword_spam / $Nspam" | bc'
Pword_ham='echo "scale=4; $Nword_ham / $Nham" | bc'
echo "estimated P($word|spam) =" $Pword_spam
echo "estimated P($word|ham) =" $Pword_ham

### PART 3
Pspam_word='echo "scale=4; $Pword_spam*$Pspam" | bc'
Pham_word='echo "scale=4; $Pword_ham*$Pham" | bc'
Pword='echo "scale=4; $Pspam_word+$Pham_word" | bc'
Pspam_word='echo "scale=4; $Pspam_word / $Pword" | bc'
echo
echo "P(spam|$word) =" $Pspam_word
cd ..
```

but the script signals that nothing that occurring in probabilization is intrinsically mysterious, elusive, or indeed particularly abstract.[11] On the contrary, the power of classifiers operates through the accumulated counting, adding, multiplying (i.e., repeated adding), and dividing (i.e., multiplying by parts or fractions) constrained by the joint probability distribution. Probability redistributes things such as emails or documents as, in this case, events in a population of words. The Naive Bayes classifies endows every word in the `Enron` dataset with a probability density function. The classification of each email becomes a matter of estimating a conditional probability based on the joint probability distribution that quantifies the chance of all the words in that email appearing together. Probabilities are often expressed as numbers between `0` and `1`, and classification entails selecting a cutoff point somewhere in this range. For instance, outputs greater than `0.5` might result in a classification as `spam`. In the `enron` dataset, "finance" has a 0.69 chance of being spam, whereas "sexy" has a chance of 1. Ironically, like the Naive Bayes classifier's own reliance on 17th- and 18th-century probability calculus, the frequent application of this machine learner to document classification and

(basic calculator), `grep` (text search using pattern matching), and `echo` (display a line of text). These tools or utilities are readily available in almost any UNIX-based operating system (e.g., Linux, MacOS, etc.). The point of using only these utilities is to illustrate the simplicity of the algorithmic implementation of the model. The first part of the code downloads the sample dataset of Enron emails (and I will discuss spam emails and their role in machine learning below). Note that this dataset has already been divided into two classes—"spam" and "ham"—and emails of each class have been placed in separate directories or folders as individual text files.

11 After fetching the dataset from a website, the code excerpted in 5.1 counts the number of emails in each category `spam` or `ham` and then counts the number of times that the chosen word (e.g., "finance" or "deal") occurs in both the spam and non-spam or ham categories. In part 2, using these counts the script estimates probabilities of any email being spam or ham and then, given that email is spam or ham, that the particular word occurs. (To estimate a probability means, in this case, to divide the word count for the chosen word by the count of the number of spam emails, and ditto for the ham emails.) In part 3, the final transformation of the data, these probabilities are used to calculate the probability of any one email being spam given the presence of that word. Again, the mathematical operations here are no more complicated than adding, multiplying, and dividing. The probability that the chosen word is a spam word is, for instance, the probability of occurrence of the word in a spam email multiplied by the overall probability that an email is spam. Finally, given that the probability of the chosen word occurring in the email dataset is the probability of it occurring in spam plus the probability of it occurring in ham, the overall probability that an email in the Enron data is spam given the presence of that word can be calculated. (It is the probability that the chosen word is a spam word divided by the probability of that word in general.)

retrieval echoes the 17th-century thinking that first conceived of the very notion of "probability" in relation to the evidential weight of documents (Hacking 1975, 85).

The improbable success of the Naive Bayes classifier

Something quite artificial is at work in the construction of these populations and their associated probability distributions. They are intentionally artificial and limited. They do not correspond or refer directly to what we know, for instance, of how language works but instead to a rather different set of concerns. Like most machine learning techniques encountering complex realities, classifiers such as Naive Bayes ignore many obvious structural or semiotic features of emails as documents (e.g., word order or co-occurrences of words). Yet this artificiality or limitation in their reference to the world allows machine learners to appear in many different guises. Despite their simple architecture, Naive Bayes classifiers have been surprisingly successful. Many machine learners transform vectorized data into probability distributions populated by fields of random variables in the process of change. They render all things as populations.

The Naive Bayes classifier attests, I am suggesting, to the altered relation between modern statistical and machine learning practice. From the early 1990s statisticians begin to generalize and rediagram Naive Bayes by examining its statistical properties more carefully. Table 5.2 shows 30 of the most cited Naive Bayes-related scientific publications.[12] The list of titles sketches a double movement. On the one hand, we see the typical diagonal forms of accumulation or positivity of a machine learner across disciplines: computer science, statistics, molecular biology (especially of cancer), software engineering, internet portal construction, sentiment classification, and image "keypoint" recognition. On the other hand, highly cited papers such as Friedman (1997) and Hand and Yu (2001) point to an intensified statistical treatment of machine learners during these years, an intensified probabilization of machine learners that strongly affects their ongoing development (leading, e.g., to the much more document-oriented, heavily probabilistic topic models appearing in the following decade; Blei, Ng, and Jordan 2003).

In *Elements of Statistical Learning*, Hastie, Tibshirani, and Friedman characterize the Naive Bayes classifier in terms of its capacity to deal with high-dimensional data:

12 Citation counts, even from the more reliable Reuters-Thomson Web of Science database, are difficult to evaluate when moving between disciplines. Some fields, such as computer science and biology, publish huge numbers of papers compared with smaller disciplines, such as astronomy or plant ecology.

Table 5.2
Most cited Naive Bayes publications 1945–2015

	Year	Title
867	2002	On Discriminative vs. Generative classifiers: A comparison of logistic regression and naive Bayes
129	2004	Molecular similarity searching using atom environments, information-based feature selection, and a naive Bayesian classifier
863	2004	Some theory for Fisher's linear discriminant function, 'naive Bayes', and some alternatives when there are many more variables than observations
297	2004	Enrichment of extremely noisy high-throughput screening data using a naive Bayes classifier
868	2004	Augmenting naive Bayes classifiers with statistical language models
870	2004	Combination of a naive Bayes classifier with consensus scoring improves enrichment of high-throughput docking results
296	2005	Not so naive Bayes: Aggregating one-dependence estimators
864	2006	Prediction of protein homo-oligomer types by pseudo amino acid composition: Approached with an improved feature extraction and Naive Bayes Feature Fusion
133	2006	Combining multi-species genomic data for microRNA identification using a Naive Bayes classifier
293	2006	Enrichment of high-throughput screening data with increasing levels of noise using support vector machines, recursive partitioning, and Laplacian-modified naive Bayesian classifiers
121	2008	Ligand-Target Prediction Using Winnow and Naive Bayesian Algorithms and the Implications of Overall Performance Statistics
223	2009	Feature selection for text classification with Naive Bayes

It is especially appropriate when the dimension p of the feature space is high, making density estimation unattractive. The naive Bayes model assumes that given a class $G = j$, the features X_k are independent (Hastie, Tibshirani, and Friedman 2009, 211).

Similar formulations can be found in most of the machine learning books and instructional materials. This appropriateness relates directly to $\forall X$ and the expansion of the vector space. As we saw in equation 5.3, p stands for the number of different dimensions or variables in the dataset. In the spam classifier, the number of dimensions balloons into hundreds of thousands because every unique word adds a new dimension to the vector space. Compared with the complications of logistic regression, neural networks or support vector machines, equation 5.3 seems incredibly simple. How is it that a simple multiplication of probabilities and the assumption that "features... are independent" can, as Hastie and co-authors write, "often outperform far more sophisticated alternatives" (Hastie, Tibshirani, and Friedman 2009, 211)?

The answer to this conundrum of success does not lie in the increasing availability of data on which to train machine learners. I want to explore two other contrasts as ways of viewing the probabilizing processes at work in Naive Bayes. The first way to view this success is in terms of *ancestral communities* of probabilization. The second concerns the statistical decomposition of machine learners in terms of their sources of error.

Ancestral probabilities in documents: Inference and prediction

Why is the Naive Bayes classifier almost always demonstrated on the problem of filtering spam email (Conway and White 2012; Schutt and O'Neil 2013, 93–113; Kirk 2014, 53; Lantz 2013, 92–93; Flach 2012; Ng 2008c), and in particular dealing with the abundance of spam emails mentioning a drug for erectile dysfunction sold under the tradename Viagra (a drug that was the byproduct of the clinical trial for hypertension and heart disease)? What are we to make of this regularity in production of statements? Admittedly spam, and spam trying to sell Viagra in particular, has been a familiar part of most email since 1997, when Viagra was approved for sale, and of all the documents that machine learners mundanely encounter in quantity in those years, email might be the most numerous as well as one of the mundanely shared. Naive Bayes classifiers and variations of them also became practical devices in managing email traffic for most people, whether they know it or not, during the mid-1990s (see, e.g., SpamAssassin). The other would be scientific publications. Many more recent machine learners train as classifiers on scientific publications (Blei and Lafferty 2007).

From an archaeological standpoint, the reiteration of email spam filtering using Naive Bayes is the effect of another process akin to the attribution of probability distributions to populations in the 19th century. Like many machine learners, Naive Bayes has one important lineage derived from the problem of classifying and retrieving documents amid archives. The operational practice of document classification is specified in the element of the archive. Genealogical affiliation with a particular problem such as document classification (or image recognition) generates many reiterations and versions of machine learners over time. As Lucy Suchman and Randall Trigg wrote in their study of work on artificial intelligence,

> rather than beginning with documented instances of situated inference... researchers begin with... postulates and problems handed down by the ancestral communities of computer science, systems engineering, philosophical logic, and the like (Suchman and Trigg 1992, 174).

Although Bayes Theorem dates from the 18th century, the highly successive use of Naive Bayes classifiers in email spam filtering in recent decades effectively draws on an ancestral community of document classification and information retrieval methods reaching back to the mid-20th century.[13]

Early attempts to use what is now called Naive Bayes in the early 1960s reiterated engagements with the evidential weight of documents that accompanied the emergence of probabilistic thinking as a quantification of belief in the 17th century (Hacking 1975, 35–49). Working at the RAND Corporation in the early 1960s, M.E. Maron described how "automatic indexing" of documents—Maron used papers published in computer engineering journals—could become "probabilistic automatic indexing." The necessary statistical assumption was to see words as populations:

> The fundamental thesis says, in effect, that statistics on kind, frequency, location, order, etc., of selected words are adequate to make reasonably good predictions about the subject matter of documents containing those words (Maron 1961, 406).

This thesis has remained somewhat fundamental in text classification and information retrieval applications, as well as many other machine learning approaches since. Maron's work focused on a collection of several hundred abstracts of papers published in the March and June 1959 issues of the *IRE Transactions on Electronic Computers*. As in contemporary supervised learning, these abstracts were divided into two groups, a training and a test set (group 1 and group 2 in Maron's terminology (Maron 1961, 407)), and the training set was classified according to 32 different categories that had already been in use by the Professional Group on Electronic Computers, the publishers of the *IRE Transactions*. Given these classifications, word counts for all distinct words in the abstracts were made, the most common terms ("the," "is," "of," "machine," "data," "computer") and the most uncommon words were removed, and the remaining set of around 1000 words were actually used for classification.

13 The other lineage descends from medical diagnosis. For instance, starting in 1960, Homer Warner, Alan Toronto, and George Veasy, working at the University of Utah and Latter-day Saints Hospital in Salt Lake City, began to develop a probabilistic computer model for diagnosis of heart disease (Warner et al. 1961; Warner, Toronto, and Veasy 1964). Their model used exactly the same "equation of conditional probability" we see in equation 5.3 but now used to "express the logical process used by a clinician in making a diagnosis based on clinical data" (Warner et al. 1961, 177). Despite the mention of logic in this description, the diagnostic model was thoroughly probabilistic in the sense that the model has no representation of logic included in its workings. Rather it calculates the probability of a given type of heart disease given "statistical data on the incidence of symptoms" (Warner, Toronto, and Veasy 1964, 558). Somewhat ironically, as they point out, physicians involved in preparing and submitting data to the diagnostic program improved the accuracy in their own diagnoses. In 1964, N.J Bailey was taking the same approach to medical diagnosis (Bailey 1965). Heart disease is a central topic in machine learning (see chapter 4 for discussion of the `South African Heart Disease` dataset).

This treatment of the abstracts as documents, then as lists of words, then as frequencies of terms, and finally as a filtered list of information-rich terms continues in much document and text classification work today. A typical contemporary information retrieval textbook such as Manning, Raghavan, and Schüijtze (2008) devotes a chapter to the topic, including the canonical discussion of how simplifying assumptions about language and meaning do not vitiate the Naive Bayes classifier. Whenever machine learners announce the unlikely efficacy of classifiers, we might attend to the ways in which previous "ancestral probabilizations" and archival constitution of the domain in question prepare the ground for that success.

Statistical decompositions: Bias, variance, and observed errors

Even with an eye on the ancestral communities that constantly accompany and steer machine learners in the world, we still need a way of accounting for the artificiality of Naive Bayes. The classifiers generates highly arbitrary probabilities of document class membership, yet these arbitrary probabilities still allow effective classification. Machine learners view the persistence of manifest artifice (in the case of Naive Bayes, a model that eschews any modeling of relations between things in the word such as words) in terms of another of the structuring differences of machine learning: the so-called *bias-variance decomposition* (Hastie, Tibshirani, and Friedman 2009, 24).

The terms "bias" and "variance" stem from the long history of statistical interest in errors (as Hacking's account of the transposition of measurement errors into population norms illustrates). The bias and variance of estimators—the estimates of the parameters of the models usually written as $\hat{\beta}$ or $\hat{\theta}$—feature heavily in machine learning discussions of prediction errors. The terms point to tensions that all machine learners experience. On the one hand, *variance* confirms the inevitable reliance of a machine learner on the data it "learns." To put it more formally, "variance refers to the amount by which $\hat{f}$ would change if we estimated it using a different training data set" (James et al. 2013, 34). On the other hand, *bias* "refers to the error that is introduced by approximating a real-life problem, which may be extremely complicated, by a much simpler model" (James et al. 2013, 35).

These two sources of error, one that results from sampling and the other arising from the structure of the model or approximating function, can be reduced or at least subject to trade-off in what *Elements of Statistical Learning* terms "the bias-variance decomposition" (Hastie, Tibshirani, and Friedman 2009, 223).[14] From the standpoint

14 Another source of error, the "irreducible error" (Hastie, Tibshirani, and Friedman 2009, 37), is noise that no model can eliminate.

of the bias-variance decomposition, every machine learner makes a trade-off between the errors deriving from differences between samples and errors due to the difference between the approximating function and the actual process that generated the data. Note that both sources of error in the bias-variance decomposition derive from transformations of the data. Variance affects how the model encounters the world (as a set of small samples or as, at the other end, a massive $N = \forall X$ dataset). Bias relating to how the model "apprehends" the data (as a set of almost coin toss-like independent events, as a geometrical problem of finding a line or curve that runs through a cloud of points, etc.).

Even with all the data, machine learning cannot fully circumvent the tensions between the different errors at work in the bias-variance decomposition. Yet sources of error do not always prove harmful. The success of Naive Bayes (and *k* nearest neighbors classifier) runs counter to the longstanding trend in statistics to construct increasingly sophisticated models of the domains they encounter. Writing in 1997, Jerome Friedman describes how simple classifiers perform surprisingly well:

> Certain types of (very high) bias can be canceled by low variance to produce accurate classification (Friedman 1997, 55).

A rather elaborate set of concepts and techniques address the bias-variance decomposition in the context of data availability. These techniques focus on managing the *test* or *generalization* error, the difference between the actual and predicted values produced by the machine learner when it encounters a fresh, hitherto unseen data sample. Machine learners in such settings still encounter the bias-variance trade-off as they select some data for training and some for testing. This trade-off has to deal with the fact that training errors—the observed difference between what the model predicts and what the training data actually shows—are not always a good guide to test generalization error. The process of fitting a model or finding a function (see previous chapter) will tend to reduce the training error by fitting the function more and more closely to the shape of the training data, but when it encounters fresh data, that function might no longer fit well. In other words, a more sophisticated function may well reduce the bias but increase the variance. "Richer collections of models" (Hastie, Tibshirani, and Friedman 2009, 224) reduce bias but tend to increase variance. Conversely, models that cope well with fresh data (and Naive Bayes is a good example of such a machine learner) display low variance but high bias.

The trade-offs between bias and variance shift markedly between different types of models and generate many different conceptual analyses of error in machine learning literature ("optimism of the training error rate" (228), "estimates of in-sample prediction error" (230), "Bayesian information criterion" (233), "Vapnik-Chervonenkis dimension" (237), "minimum description length" (235)) and technical methods of estimating prediction error ("cross-validation" (241), "bootstrap methods" (249), "expectation-maximization algorithm" (272), "bagging" (282), or "Markov Chain Monte Carlo (MCMC)" (279)), many of which date from the 1970s (e.g., cross-validation (Stone 1974), bootstrap (Efron 1979), and expectation-maximization (Dempster, Laird, and Rubin 1977)).

A daunting field of concepts, themes, techniques, and methods all gravitate to the threshold of probabilization. They invoke in some cases sophisticated mathematical or statistical constructs. They also often rely on computational iteration or infrastructural scale to optimize parameters in models whose underlying intuitions remain quite straightforward (as in a linear regression or Naive Bayes). In some cases, the implementation of a model may be simple, but analysis of how the machine learner manages to curtail a source of error such as bias or variance entails much more sophisticated statistical understanding. Many analyses of how a model becomes a useful approximation reconfigure the models as members of a population whose variations and uncertainties, whose tendencies and predispositions must be sampled, tested, and monitored. The bias-variance decomposition points to an irreducible friction in the way that machine learning structures differences in the world.

Does machine learning construct a new statistical reality?

Following a broadly Foucauldian line of argument, Hacking proposes that statistical thinking and practice in the 19th and early 20th century ontologically reconfigured things in terms of probability distributions (and the Gaussian distribution in particular). What happens in worlds where the statistical treatment of error—the bias-variance decomposition is a shorthand term for this—distributes probability throughout an operational formation? I have suggested that an ancestral probabilization of domains and the statistical decomposition of error come together in statistical machine learning. The bias-variance decomposition includes both tightly bound points and certainly relatively free or unbound points, as we saw in the case of the Naive Bayes classifier in its encounter with data. It generates highly erroneous probability estimates but performs well as a classifier.

Viewed diagrammatically, unbound points matter greatly to the relations of force at work in a knowledge-power conjunction. Probabilization gives machine learning a relation to its own plurality, to the tendencies of its models to proliferate and vary. Every attempt to construct a machine learner in a given setting draws on both the reiteration of ancestral probabilities (i.e., prior structuring of settings in conformity with some probability distribution) or the many interactive adjustments, redistributions, and resamplings of the data *and* transformations of the models associated with the bias-variance decomposition.

Mayer-Schönberger and Cukier argue that having much data or all data ($N = \forall X$) alters knowledge. Versions of this claim can be found running through various scientific and business settings throughout the 20th century.[15] In certain settings, $N = all$ has been around for quite a while (as, e.g., in many document classification settings where the whole archive or corpus of documents has been electronically curated for decades). Mayer-Schönberger and Cukier rightly emphasize that the huge quantities of data sluicing through some contemporary infrastructures support wider inferences (11). Their discounting of statistical sampling as a concept "developed to solve a particular problem at a particular moment in time under specific technological constraints" (Mayer-Schönberger and Cukier 2013, 31) does not, however, accommodate the operational practices of sampling that pervade machine learning, particularly in the forms of probabilization.

Whether someone uses Naive Bayes, a topic model, neural networks, or logistic regression does not greatly alter the processes of probabilization. Random variables, probability distributions, errors, and model selection practices crowd in around and reconfigure machine learners as members of a statement-generating population. In many ways, the Mayer-Schönberger and Cukier account bobs in the wake of the enterprise-wide accumulations of data. They pay so much attention to the capital potentials of data accumulation that they cannot easily attend to the prior question of how machine learners probabilize those data. Sampling, estimation, likelihoods, and a whole gamut of dynamic relationships between random variables in joint probability distributions reassert themselves amid a population of models. The data may not be sampled, but models moving through the high-dimensional vector spaces opened up by having all the data transform it probabilistically. Despite the fact that not all machine learners are strictly speaking probabilistic models,[16] machine learners relate to themselves and the data as populations defined by probability distributions.

15 Later chapters of this book will track several instances of having all the data in the sciences, government, and business to show what having all the data entails in different settings.

16 Machine learning textbooks written by computer scientists tend to define probabilistic models more narrowly. As Peter Flach suggests:

> Probabilistic models view learning as a process of reducing uncertainty using data. For instance, a Bayesian classifier models the posterior distribution $P(Y|X)$ (or its counterpart, the likelihood function $P(X|Y)$) which tells me the class distribution Y after observing the features values X (Flach 2012, 47).

But regardless of whether they are probabilistic in this sense, the evaluation and configuring of machine learners irreducibly depends on a statistical treatment of errors and their trade-offs.

Machine learning inhabits a reality that had already introjected statistical realities at least a century earlier, whether through the social physics of Quetelet, the biopolitical norms of Francis Galton and his regression to the mean (the linear model of regression is probably the basic machine learning model), or later, in the probability functions of quantum mechanics in early 20th-century physics. Assembling an aggregate reality of many devices, machine learning inverts probability distributions. In this inversion, probability distributions, which had become the operational statement and model of truth for many different kinds of populations, fold back or redistribute themselves into devices such as machine learners whose variations and uncertainties become populations. Populations of models are sampled, measured, and aggregated in the ongoing production of statistical realities whose object is no longer a property of individual members of a population (their height, life expectancy, and chance of HIV/AIDS) but members of a population of models.

6 Patterns and Differences

The notion of pattern involves the concept of different modes of togetherness (Whitehead 1956, 195–196).

Algorithms for pattern recognition were therefore from the very beginning associated with the construction of linear decision surfaces (Cortes and Vapnik 1995, 273–274).

Do machine learners generate new patterns of difference? Should we hold machine learners accountable for their claims to recognize patterns in data in the same way we hold experimental scientists accountable for their factual claims?[1] This chapter explores two major machine learning treatments of pattern dating from the last decades of the 20th century from the standpoint of differences. I suggest that what counts as pattern changes in machine learning over time. Although much machine learning strains to identify differences in terms of differences of degree, the practice of pattern finding harbors differences of kind. For critical thought, the connection between pattern and differences is particularly important because if machine learning changes what counts as pattern, then this change will also affect the recognition or articulation of differences. We have seen the emergence of the vector space and its vectorized transformations, the multiplication of operational functions and their associated partial observers, and then the probabilization that distributes machine learners into

1 Many authors have suggested that algorithms should be the focus of more attention. The sociologist Mike Savage, in his account of the growth of "descriptive assemblages" based around large-scale data mining of transactions, administrative records, and social media practice concludes:

> It follows that a core concern might be to scrutinize how pattern is derived and produced in social inscription devices, as a means of considering the robustness of such derivations, what may be left out or made invisible from them, and so forth. We need to develop an account which seeks to criticize notions of the descriptive insofar as this involves the simple generation of categories and groups, and instead focus on the fluid and intensive generation of potential (Savage 2009, 171).

populations of error-sensitive learners. What in this diagram and in the forest-like growth of techniques, projects, applications, and proponents allows us to make sense of what happens to differences in machine learning?

Across vectors, functions, and populations, the diagram of machine learning weaves and knots many points of emergence, continuity, and conjunction. I view the formidable accumulations of infrastructure, devices, and expertise accrediting around machine learning as multifaceted abstractions, where abstraction is understood diagrammatically as a concretizing entanglement of references in an operational formation. Three highly developed and heavily used machine learners—decision trees, support vector machines, and neural nets—more or less mesmerized machine learning between 1980 and 2000. They initiated relatively novel and somewhat heterogeneous diagrammatic movements into data. These diagrammatic movements, which we might characterize as *splitting* and *marginalizing*, not only animate subsequent machine learners in producing newer techniques but also reconfigure what counts as pattern. Because machine learning has no fixed idea of pattern (the term lacks an operational definition), it claims that machine learners uncovering hidden patterns in data might be better grounded in the operational practices of working with differences.[2]

Like other machine learners of recent decades, the decision tree and support vector machine embody an enunciative modality, a way of describing, locating, and perceiving differences in which differences of degree and kind are remapped. Every machine learner generates statements, but from different places, by somewhat different individuals, and from the different situations they "occupy in relation to the various domains or groups of objects" (Foucault 1972, 52).[3] Practically, decision trees and support vector machines loom large in various contemporary accounts of machine learning as a way of knowing (e.g., in popular machine learning books such as *Machine Learning for Hackers* (Conway and White 2012) or *Doing Data Science* (Schutt and O'Neil 2013)). The machine learning research published in statistics, computer science, mathematics,

2 *Elements of Statistical Learning* uses the term "pattern" only occasionally. The term appears 33 times there and mainly in the bibliography. Apart from Brian Ripley's *Pattern Recognition and Neural Networks* (Ripley 1996), statisticians largely eschew the term. Computer scientists like it more, particularly in work on the classification of images (see Christopher Bishop *Pattern Recognition and Machine Learning* (Bishop 2006). Hastie, Tibshirani, and Friedman, as statistical machine learners, confine their use of pattern to the term "pattern recognition."

3 Although Foucault tends to retain a decoupled subject-object relation in the production of statements, I tend to see these enunciative modalities as distributed across people and things. As always, machine learner is a composite term for this distribution.

artificial intelligence, and a swathe of related scientific fields from 1980 to 2010 bristles with references to decision trees and support vector machine, as well as neural networks.[4]

Rather than seeing pattern as something discovered in data, the notion of enunciative modality suggests that we should examine the diagrammatic operations that configure differences in the practice of machine learning, giving rise to a field of patterns attributed to objects or subject positions. Even if they are not the latest inventions, the two machine learners that anchor this chapter are perhaps the most distinctive data mining, pattern recognition, and predictive modeling achievements of the late 20th century (at least judging by the citations and usage they attract). They differ greatly in how they move through data. At certain times, they come together (e.g., in machine learning competitions discussed in chapter 8; in certain formalizations, such as machine learning theory or in graphs of the bias-variance decomposition discussed in chapter 5; or in the pedagogy of machine learning discussed in chapter 2), but my focus is on their differences.

Splitting and the growth of trees

Mastering the details of tree growth and management is an excellent way to understand the activities of learning machines generally (Malley, Malley, and Pajevic 2011, 118).

Decision trees promise an understanding of machine learning. The enunciative modality of the decision tree concerns the observability and comprehensibility of machine learning. As we have seen and will see, not all machine learners readily support observation or comprehension. The cost of decision tree comprehensibility, however, is a certain highly restricted framing of differences. As *Elements of Statistical Learning* puts it, "tree-based methods partition the feature space into a set of rectangles and then fit a simple model (like a constant) in each one. They are conceptually simple yet powerful" (Hastie, Tibshirani, and Friedman 2009, 305).

4 The top 20 most cited publications in the field include Ross Quinlan and Leo Breiman's papers on decision trees (Quinlan 1986; Breiman et al. 1984), Vladimir Vapnik and Corinna Cortes' support vector machines papers (Vapnik 1999; Cortes and Vapnik 1995), an early textbook written by a computer scientist on machine learning (Mitchell 1997), a textbook and software package on data mining using Java (Witten and Frank 2005), a textbook on pattern recognition dating from the 1970s (Duda, Hart, and Stork 2012), a tutorial on an error control technique (ROC—Receiver Operating Characteristics, first developed by the U.S. military during WWII), and somewhat lower, another well-known textbook, this time on neural networks and pattern recognition (Bishop 2006).

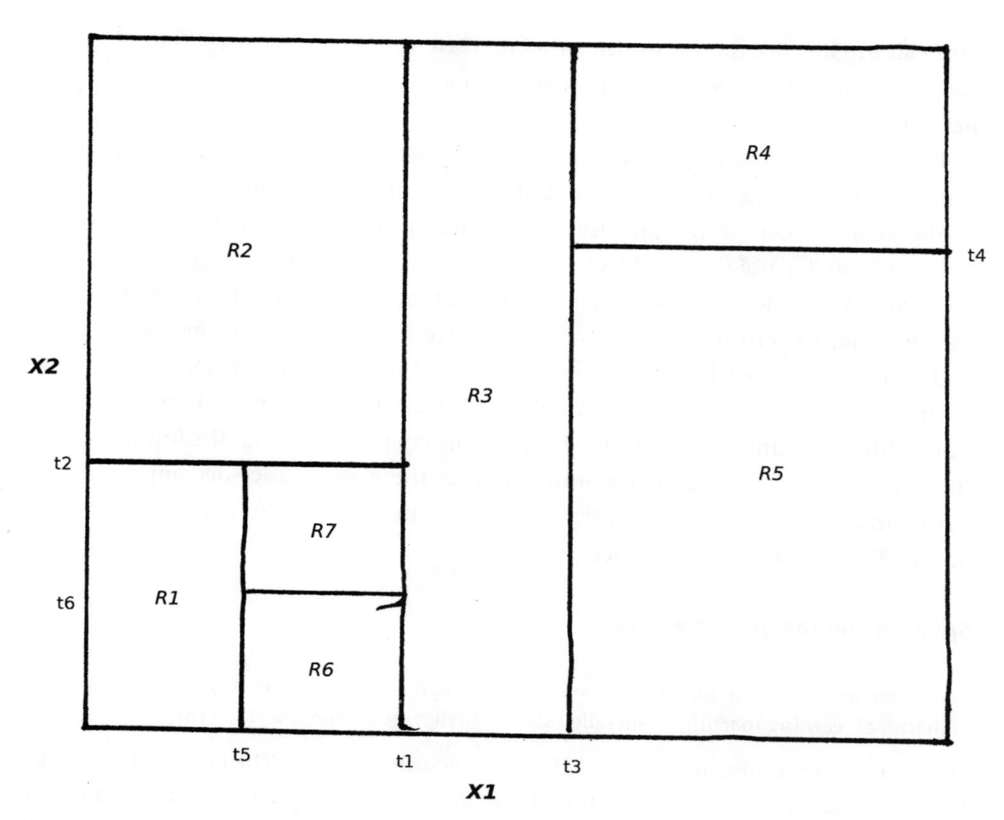

Figure 6.1
Recursive partitioning of the feature space.

Tree-based methods are supervised learners as they require some training data either labeled by class or some other outcome value. The variable types in the feature space (or vector space of the data) can be mixed. Because the method cuts the vector space into a tiled surface (see figure 6.1), the features or data variables can be continuous or discontinuous. The simple models that tree methods construct each define one of the rectangular regions or partitions of the feature space. In figure 6.1, the different regions or partitions produced by a decision tree are labeled *R*1, *R*2, and so on.

Work on classification and regression techniques using decision trees goes back to the early 1960s, when social scientists James Morgan and John Sonquist at the University of Michigan's Institute for Social Research were attempting to analyze increasingly large social survey datasets (Morgan and Sonquist 1963). As Dan Steinberg describes in his brief history of decision trees (Steinberg and Colla 2009, 180), the "automatic

Table 6.1
References to Morgan and Sonquist's Automatic Interaction Detector

Title	Year	Citations
The Achievement Motive And Economic Behavior	1964	20
Simplification Of Economic Models	1966	9
Data Dredging Procedures In Survey Analysis	1966	65
Advertising Performance As A Function Of Print Ad Characteristics	1967	32
World Affairs Information And Mass Media Exposure	1967	27
Juvenile Probation System Simulation For Research And Decision Making	1968	16
Presidential Elections Explanation Of Voting Defection	1969	12
An Interactive Technique For Analysis Of Multivariate Data	1969	9
Finding Variables That Work	1969	22
Brand Trial After A Credibility Change	1970	7

interaction detector" (AID), as it was known, sought to automate the practice of data analysts looking for interactions between different variables. The variety and sheer optimism of subsequent applications of these prototype decision tree techniques is striking. In the 1960s and 1970s, papers that drew on the AID paper or use AID techniques can be found, as table 6.1 shows, in education, politics, economics, population control, advertising, mass media, and family planning.

A decade after the initial work, AID was the object of trenchant criticism by statisticians and others, not for the classifications it used (see figure 6.2) but for its pure empiricism. Writing in the 1970s, statisticians in the behavioral sciences, such as Hillel Einhorn at the University of Chicago, castigated the use of such techniques. The criticisms stemmed from a general distrust of "purely empirical methods" and scepticism focused on their positivity:

> The purely empirical approach is particularly dangerous in an age when computers and packaged programs are readily available, since there is temptation to substitute immediate empirical analysis for more analytic thought and theory building. It is also probably too much to hope that a majority of researchers will take the time to find out how and why a particular program works. The chief interest will continue to be in the output-the results-with as little delay as possible (Einhorn 1972, 368).

Einhorn discusses AID alongside other techniques, such as factor analysis and multidimensional scaling (both still widely used), before concluding, "it should be clear that proceeding without a theory and with powerful data analytic techniques can lead to large numbers of Type I errors" (Einhorn 1972, 378). His statistical objections to AID are particularly focused on the problematic power of the technique: "it may make sense out of 'noise'" (369). Consequently, researchers easily misuse the

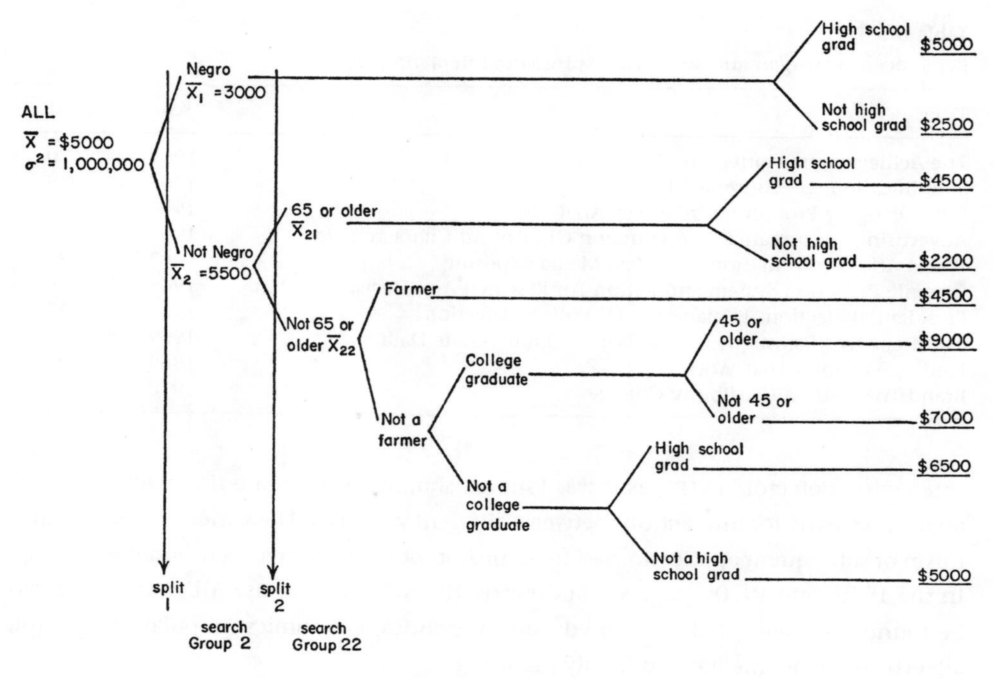

Figure 6.2
AID classifies annual earnings (Morgan & Sonquist, 1963, 430).

technique: they "overfit" the data and do not pay enough attention to issues of validation. Similarly, the British marketing researcher Peter Doyle, criticizing the use of AID in assessing store performance and site selection by operations researchers, complained that searching for patterns in data using datasets was bound to lead to spurious results, and the decision trees, although intuitively appealing (i.e., they could be easily interpreted), were afflicted with arbitrariness: "a second variable may be almost as discriminating as the one chosen, but if the program is made to split on this, quite a different tree occurs" (Doyle 1973, 465–466).[5] Most of these criticisms can be seen as expressing conventional statistical caution in response to threats to validity, but they also address the core issues of pattern and difference: did the trees render differences arbitrarily?

5 These objections and resistances to early decision trees echo today in discussions around pattern recognition, knowledge discovery, and data mining in science and commerce. The problem of what computers do to the analysis of empirical data is longstanding.

1984: Differences in recursive partitioning

As Einhorn expected, it was too much to hope that all researchers would take time to investigate how a particular program works. Statistical researchers, however, did take time in the following decade to investigate how decision trees work. Writing around 2000, Hastie, Tibshirani, and Friedman, who could hardly by accused of not understanding decision trees, happily recommend decision trees as the best off-the-shelf classifier: "of all the well-known learning methods, decision trees comes closest to meeting the requirements for serving as an off-the-shelf procedure for data-mining" (Hastie, Tibshirani, and Friedman 2009, 352). We might wonder here, however, whether they damn with faint praise because "off-the-shelf" suggests prepackaged and commodified, and the term "data-mining" is not without negative connotations. As for its commercial realities, in 2013, Salford Systems, the purveyors of the leading contemporary commercial decision tree software, CART, could claim:

> CART is the ultimate classification tree that has revolutionized the entire field of advanced analytics and inaugurated the current era of data mining. CART, which is continually being improved, is one of the most important tools in modern data mining. Others have tried to copy CART but no one has succeeded as evidenced by unmatched accuracy, performance, feature set, built-in automation and ease of use.Salford Systems

What happened between 1973 and 2013? Decision trees somehow stepped out of the statistically murky waters of social science departments and business schools in the early 1970s to inaugurate the "current era of data mining" (which the scientific literature indicates starts in the early 1990s). This innovation was not only commercial. As the earlier citation from U.S. National Institutes of Health biostatisticians Malley, Malley, and Pajevic indicates, decision trees enjoy high regard even in biomedical research, a setting where statistical rigor is highly valued for life-and-death reasons. The happy situation of decision trees four decades on suggests that some kind of threshold was crossed in which the epistemological, statistical, or algorithmic ("built-in automation") power of the technique altered substantially.

The third author of *Elements of Statistical Learning*, Jerome Friedman, worked at the U.S. Department of Energy's Stanford Linear Accelerator during the late 1970s. Friedman was instrumental in rescuing decision trees from the ignominy of profligate ease of use and pure empiricism they had endured since the late 1960s. The reorganization and statistical retrofitting of the decision tree was not a single or focused effort. During the 1980s, statisticians such as Friedman and Leo Breiman renovated the decision tree as a statistical tool (Breiman et al. 1984). At the same time, computer scientists such as Ross Quinlan in Sydney were reimplementing decision trees guided by an artificial intelligence-based formalization as rule-based induction technique (Quinlan 1986).[6] This uneasy parallel effort between computer science and statistics still somewhat strains relations in machine learning today. Statisticians and computer scientists do use the same techniques but often with the computer scientists focusing on optimization and algorithmic scale and the statisticians inventing novel statistical formalizations and abstractions. The fateful embrace of statistics and computer science, the disciplinary binary that vectorizes machine learning, has been generative in the retrieval of the decision tree.

An initial symptom of the transformation of the technique appears in a name change. The term "decision tree," although still widely used in the research literature and machine learner parlance, was supplanted by "classification and regression tree" during the late 1970s and 1980s. The more statistical-sounding "classification" and "regression tree" are sometimes contracted to CART, a term that refers to a computer program described in Breiman et al. (1984) as well as the title of that highly cited monograph, *Classification and Regression Trees*. As we have seen in previous chapters, classification and regression (predictive modeling using estimates of relations between variables) refer to perhaps the two main types of machine learning practice. Their concatenation with "tree" attests to a renovation of existing machine learning approaches behind a single facade.

The implementation of machine learning techniques in `R` accentuates the statistical side of decision tree practice, but that has certain forensic virtues not offered by commercial or closed-source software often produced by data miners. The name of one longstanding and widely used `R` package attests to something: `rpart` is a contraction of "recursive partitioning," and this term generally describes how the decision tree algorithm works to partition the vector space into the form shown in figure 6.1 (Therneau, Atkinson, and Ripley 2015). CART, in contrast, is a registered trademark of Salford Systems, the software company mentioned earlier, that sells the leading commercial implementation of classification and regression trees. Hence, the `R` package `rpart`

6 Quinlan's papers and book on versions of the decision tree (`ID3` and `c4.5`) are both among the top 10 most highly cited references in the machine literature. Google Scholar reports more than 20,000 citations of the Quinlan's book *C4.5: Programs for Machine Learning* (Quinlan 1993) (although far fewer appear in Thomson Reuters Web of Science). Several years ago, `C4.5` was voted the top data mining algorithm (Wu et al. 2008). Although I don't discuss Quinlan's work in much detail here, we should note as a computer scientist, Quinlan takes a much more rule-based approach to decision tree than Breiman and co-authors.

Listing 6.1
Decision tree for `iris` dataset

```
data(iris)
library(rpart)
iris_tree =rpart(Species ~ ., iris)
```

cannot call itself the more obvious name `cart` and instead invokes the underlying algorithmic process: recursive partitioning.[7]

R.A. Fisher's `iris` dataset, which contains 150 measurements made in the 1930s of petal and sepal lengths of *iris virginica, iris setosa*, and *iris versicolor*, is a standard instructional example for decision trees (Fisher 1938).[8] The code shown here loads the `iris` data (the dataset is routinely installed with many data analysis tools), loads the `rpart` decision tree library, and builds a decision to classify the irises by species. What has happened to the iris data in this decision tree? The `R` code that invokes the recursive partitioning algorithm is so brief that we can't tell much about how the data have been recursively partitioned. We know that the *iris* has 150 rows, and that there are equal numbers of the three iris varieties.

Code brevity indicates that a great deal of formalization of practice has accrued around decision trees. Some of this formalization was described in the *Classification and Regression Trees* monograph (Breiman et al. 1984). Classification in decision trees operates by splitting each dimension of the vector space into two parts (as we saw in figure 6.1). These splits institute branches along which differences are hierarchically

7 Other `R` packages such as `party` (Hothorn, Hornik, and Zeileis 2006) and `tree` (Ripley 2014) also use recursive partitioning, but with various tweaks and optimisations that I leave aside here.

8 `iris` is a small dataset, a precomputational miniature. That diminutive character makes it diagrammatically mobile. It supports a rhizomatic ecosystem of examples scattered across the machine learning literature. The usual framing of the classification problem is how to decide whether a given iris blossom is of the species *virginica, setosa*, or _versicolor. These irises don't grow in forests—they are more often found in riverbanks and meadows—but they do offer a variety of illustrations of how machine learning classifiers are brought to bear on classification problems. Here the classification problem is taxonomic: the *iris* genus has various subgenera and sections within the subgenera. Setosa, *virginica*, and *versicolor* all belong to the subgenus *Limniris*. This botanical context is routinely ignored in machine learning applications. In machine learning textbooks and tutorials, `iris` typically would be used to demonstrate how cleanly a classifier can separate the different kinds of irises.

ordered in a tree structure. The recursive splitting algorithm draws a diagram of hierarchical differences. The operational problem here is that many splits are possible. What is a good split or ordering of differences?

The first problem in tree construction is how to use $\mathcal{L}$ to determine the binary splits of $\mathcal{X}$ into smaller pieces. The fundamental idea is to select each split of a subset so that the data in each of the descendant subsets are "purer" than the data in the parent subset (Breiman et al. 1984, 23).

Tree construction hinges on the notion of purity or more precisely "node impurity," a function that measures the extent to which data labeled as belonging to different classes are mixed together at a given branch or node in a decision tree: "that is, the node impurity is largest when all classes are equally mixed together in it, and smallest when the node contains only one class" (Breiman et al. 1984, 24). As Malley and coauthors note, "the collection of purity measures is still a subject of research" (Malley, Malley, and Pajevic 2011, 123), but Breiman, Friedman, Olshen, and Stone promoted a particular form of impurity measure for classification trees known as "Gini index of diversity" (Breiman et al. 1984, 38). Like the planar decision surface used in classifiers such as the perceptron or logistic regression model, recursive partitioning combined with measures of node impurity transforms data by cuts or divides. Whereas in linear model-based machine learners, the intuition motivating the function finding or learning was "find the line that best expresses the distribution of the data," here the intuition is more like "find the cuts that minimize mixing." Good splits decrease the level of impurity in the tree. In a tree with maximum purity, each terminal node—the nodes at the base of the tree—would contain a single class.

In figure 6.3, the plot on the left shows the decision tree and the plot on the right shows just *setosa* and *versicolor* plotted by petal and sepal widths and lengths. Decision trees are read from the top down, left to right. The top level of this tree can be read, for instance, as saying, if the length of petal is less 2.45, then the iris is *setosa*. As the plot on the right shows, most of the measurements are well clustered. Only the *setosa* petal lengths and widths seem to vary widely. All the other measurements are tightly bunched. A decision tree has little trouble ordering differences between species of iris.

Like logistic regression models, neural networks, support vector machines, or any other machine learning technique, decision trees order differences in terms of specific qualities and logics. Recursive partitioning splits and subdivides the vector space to capture increasingly small differences between cases, and thereby achieves an ever-closer fit to the individual or subindividual variations. Although the partitioning or splitting rules have strong statistical justifications, they do not at all eliminate the problem of instability or variance in trees. For instance, they easily end up "overfitting" the data.

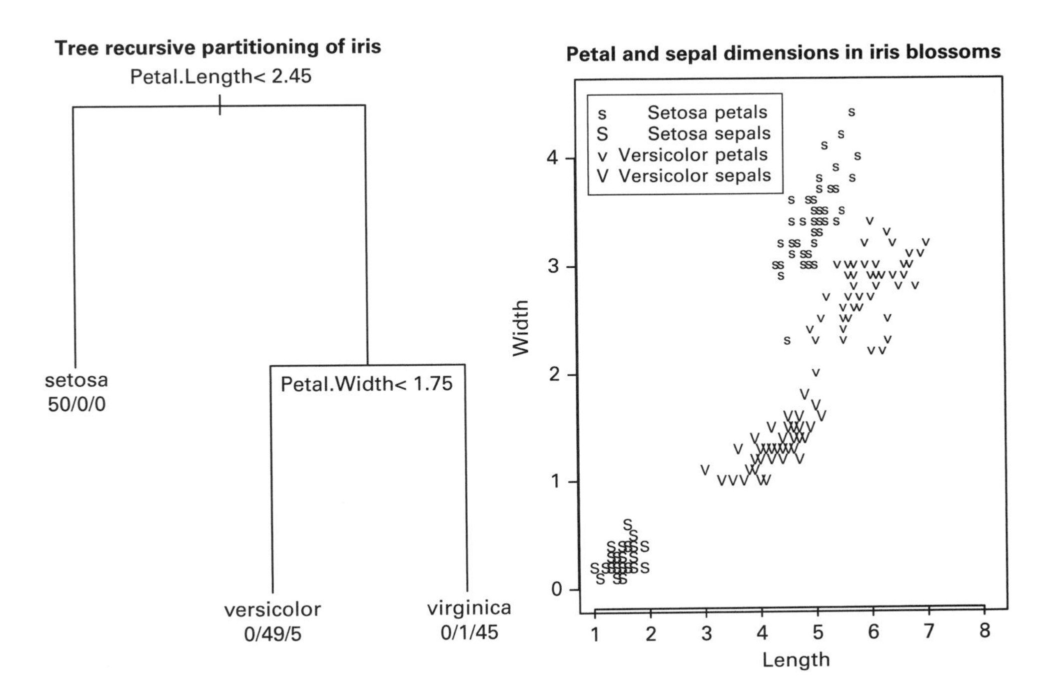

Figure 6.3
Decision tree on `iris` dataset.

Overfitting is a problem for all machine learning techniques. Algorithms sometimes find it hard to know when to stop identifying differences. During construction of a decision tree, recursive partitioning splits features in the data into smaller and smaller groups. "The goodness of the split," wrote Breiman and co-authors, "is defined to be the decrease in impurity" (Breiman et al. 1984, 25). Under this definition of goodness, the terminal nodes or leaves of the tree can, as mentioned above, end up containing a single case or a single class of cases.

The decision tree targets the differences of the individual case to such a degree that it could end up seeing categorical differences everywhere. Operating to maximize the purity of the partitions it creates, it leans too heavily on data it has been trained on to see relevant similarities when fresh data appear. Trees that branch too much are sensitive to differences and generalize poorly (i.e., they suffer from generalization). Such a model will almost always *overfit* because slight variations in the values of variables in a fresh case are likely to yield widely differing predictions. In the terminology of machine learning, such a decision tree may have low bias but high variance.

Limiting differences

Given this problem of unstable difference, much of the development of decision trees did not revolve around how to construct them but how to limit their growth so as to manage tensions between pure but unstable differences and impure but stable classification. As *Elements of Statistical Learning* puts the problem in its account of classification and regression trees:

How large should we grow the tree? Clearly a very large tree might overfit the data, while a small tree might not capture the important structure. Tree size is a tuning parameter governing the model's complexity, and the optimal tree size should be adaptively chosen from the data. One approach would be to split tree nodes only if the decrease in sum-of-squares due to the split exceeds some threshold. This strategy is too short-sighted, however, since a seemingly worthless split might lead to a very good split below it. The preferred strategy is to grow a large tree T_0, stopping the splitting process only when some minimum node size (say 5) is reached. Then this large tree is pruned using cost-complexity pruning (Hastie, Tibshirani, and Friedman 2009, 307–308).

Growing a maximum decision tree and then cutting back its branches using a cost-function optimizes the decision tree as a machine learner. "Cost complexity pruning" extends the optimization we have already discussed in relation to linear regression and logistic regression models (in chapter 4). As in the partial observers associated with these techniques, cost complexity pruning controls the "complexity" of a tree—how many branches and leaves/nodes it contains, combined with measures of how well it classifies or predicts—by iteratively observing and comparing different versions of trees with each other. "We define the cost complexity criterion," write Hastie and co-authors, as:

$$C_\alpha(T) = \sum_{m=1}^{|T|} N_m Q_m(T) + \alpha|T| \quad (6.1)$$

The idea is "to find, for each α, the subtree $T_\alpha \subseteq T_0$ to minimize $C_\alpha(T)$" (Hastie, Tibshirani, and Friedman 2009, 308). For present purposes, we need only recognize that the cost complexity function reconfigures a large decision tree (T_0 in equation 6.1) by cutting or pruning it back through optimization that balances between the complexity of the tree and its stability. Tree construction is as an optimization problem, in which the variation of a parameter (α) allows minimization of a derived value (the cost C_α).

Even if the graphic form of the decision tree was, by virtue of the longstanding diagrammatic practice of tree drawing, easy to interpret, machine learners had no way of gauging the instability or variability of any given tree. Hastie and co-authors write, "One major problem with trees is their high variance. Often a small change in the data can result in a different series of splits, making interpretation somewhat precarious. The major reason for this instability is the hierarchical nature of the process: the effect of an error in the top split is propagated down to all of the splits below it. The diagrammatic form that allows decision trees to be observed and interpreted is also the source of their instability. Regardless of this instability, the diagrammatic composition of the tree through splitting and pruning negotiates between two different ways of doing difference.

The shift from `AID` to `CART` enunciates a change in how patterns of difference become visible. The decision tree algorithm superimposes recursive partitioning and cost-complexity pruning to configure a mode of enunciation of differences. It creates a new rules of differentiation of individuals, facts, things, and relations. Differences in a decision tree—the combination of purity and density that comes from recursive partitioning and cost-complexity pruning—reconfigure what counts as pattern.

Decision trees have been heavily used in credit risk assessment as well as many biomedical models. Does their popularity stem from the legibility of the statements they produce, even if those patterns prove unstable? Or is the success of the decision tree perhaps better understood as an effect of a change in the differentiation of patterns more generally, their mode of enunciation, in which case decision trees would only be one instance among many? If we understand machine learners as generating populations of statements, the transformation and remodeling of the decision tree as classification and regression trees suggests a subtle, nonlocalizable discontinuity. The later development of the decision tree and its subsequent transmogrification into random forests (Breiman 2001b) that grow a myriad of small decision trees disperses kaleidoscopic fragments of classificatory order with only partial or provisional stabilization in visible pattern. In such developments—and we could also consider here techniques, models, and methods of "boosting," "bagging," or the "ensemble learning" that conducts "supervised search in a high-dimensional space of weak learners" (Hastie, Tibshirani, and Friedman 2009, 603)—pattern has an increasingly operational rather than a visible mode of togetherness.

The successful dispersion of the support vector machine

In growing and pruning decision trees, and even more markedly in subsequent machine learners, patterns play out in dispersion and discontinuity rather than in regular geometry. Although machine learners order differences, that ordering becomes increasingly difficult to see in its dispersion. Take the case of the support vector machine. The second most highly cited reference in the last few decades of machine learning literature is a paper from 1995 by Corinna Cortes and Vladimir Vapnik of AT&T Bell Labs in New Jersey titled "Support Vector Networks" (Cortes and Vapnik 1995). Few women's names appear prominently in the machine learning literature. The computing science and statistics departments at Stanford and Berkeley, the laboratories at Los Alamos, and AT&T Bell between the 1960s and 1980s were, it seems, not overly popular or populated with women scientists and engineers. Some prominent machine learning researchers at the time of writing are women (I return to this in chapter 8), but Cortes is perhaps preeminent both as head of Google Research in New York (2014) and as recipient with Vapnik of an Association for Computing Machine award in 2008 for work on the support vector machine algorithm.[9]

The rapid rise to popularity of the support vector machine can be seen in the machine learning research literature, a small slice of which appears in table 6.2. A substantial fraction of the overall research publication since the mid-1990s accumulates around this single technique and as usual ranges across credit analysis, land cover prediction, protein structures, brain states, and face recognition. The support vector

9 The support vector machine is distinctive in its transformations of data, and this owes something to history, politics, and geography. Vapnik trained and worked for decades in the former USSR as a mathematician and statistician. His writings on the problems of pattern recognition contrast greatly with other engineers, statisticians, and computer scientists in their robustly theoretical formalism. A highly cited 1971 publication with Alexey Chervonenkis "on the uniform convergence of relative frequencies of events to their probabilities" (published in Russian in 1968) (Vapnik and Chervonenkis 1971) sets the formal tone of this work. In ensuing publications in Russian and then in English after Vapnik moved from Moscow to AT&T's New Jersey Bell Labs in 1990, Vapnik's work remains quite formally mathematical. Although it pertains to learning machines, machine here are understood mathematically simply as "the implementation of a set of functions" (Vapnik 1999, 17). The way that Vapnik develops a theory of learning owes little visible debt to actual attempts to work with data or experience in doing statistics in any particular domain. This contrasts greatly, for instance, with the work of statisticians such as Breiman or Friedman or even computer scientists such as Quinlan or Le Cun, whose work lies much closer to fields of application. Vapnik's work, like that of the Russian mathematician Andrey Kolmogorov he draws on, differs from many other contributions to machine learning partly by virtue of this formality and its efforts to derive insight into machine learning by theorising learning. The *Vapnik-Chervonenkis dimension* (VC dimension), a widely used way of defining the capacity of a particular machine learning technique to recognize patterns in data dates from his work in the 1960s and underpins a general theory of "learning." Vapnik writes in 1995,

> The VC dimension of the set of functions (rather than the number of parameters) is responsible for the generalization ability of learning machines. This opens remarkable opportunities to overcome the "curse of dimensionality (Vapnik 1999, 83).

As we will see in this chapter, Vapnik's attempts to overcome dimensionality also reshape what counts as pattern.

Table 6.2
Most cited papers on support vector machines

Title	Year	Citations
A tutorial on Support Vector Machines for pattern recognition	1998	3510
Support vector machine classification and validation of cancer tissue samples using microarray expression data	2000	869
Choosing multiple parameters for support vector machines	2002	681
Classification of hyperspectral remote sensing images with support vector machines	2004	428
Comparing support vector machines with Gaussian kernels to radial basis function classifiers	1997	405
Support Vector Machines for 3D object recognition	1998	396
An assessment of support vector machines for land cover classification	2002	295
Drug design by machine learning: support vector machines for pharmaceutical data analysis	2001	273
A novel method of protein secondary structure prediction with high segment overlap measure: Support vector machine approach	2001	266
LIBSVM: A Library for Support Vector Machines	2011	246
The entire regularization path for the support vector machine	2004	223
A GA-based feature selection and parameters optimization for support vector machines	2006	223
The support vector machine under test	2003	219
Credit rating analysis with support vector machines and neural networks: a market comparative study	2004	218

machine spans the normal biopolitical triangle of life, labor, and language. The influence of the technique can also be seen in overlapping fields such as pattern recognition and data mining, where Cortes and Vapnik (1995) and similar papers rank near the top-cited papers.[10] This kind of growth betokens high levels of interest, identification, and investment on the part of machine learners, and presumably more widely.

I suggested earlier that the classification tree (and then random forests) illustrates an enunciative modality anchored in a tension between recursively partitioned differences and classificatory stability. The support vector machine shown in figure 6.4 demonstrates another change in what counts as pattern. The decision boundaries in the subgraphs have different contours, contours that suggest a more flexible construction. Although the name "support vector machine" is somewhat forbiddingly technical compared with more familiar terms, such as "decision tree" or even "neural network,"

10 *Elements of Statistical Learning* also devotes a chapter to support vector machines (Hastie, Tibshirani, and Friedman 2009, chapter 14).

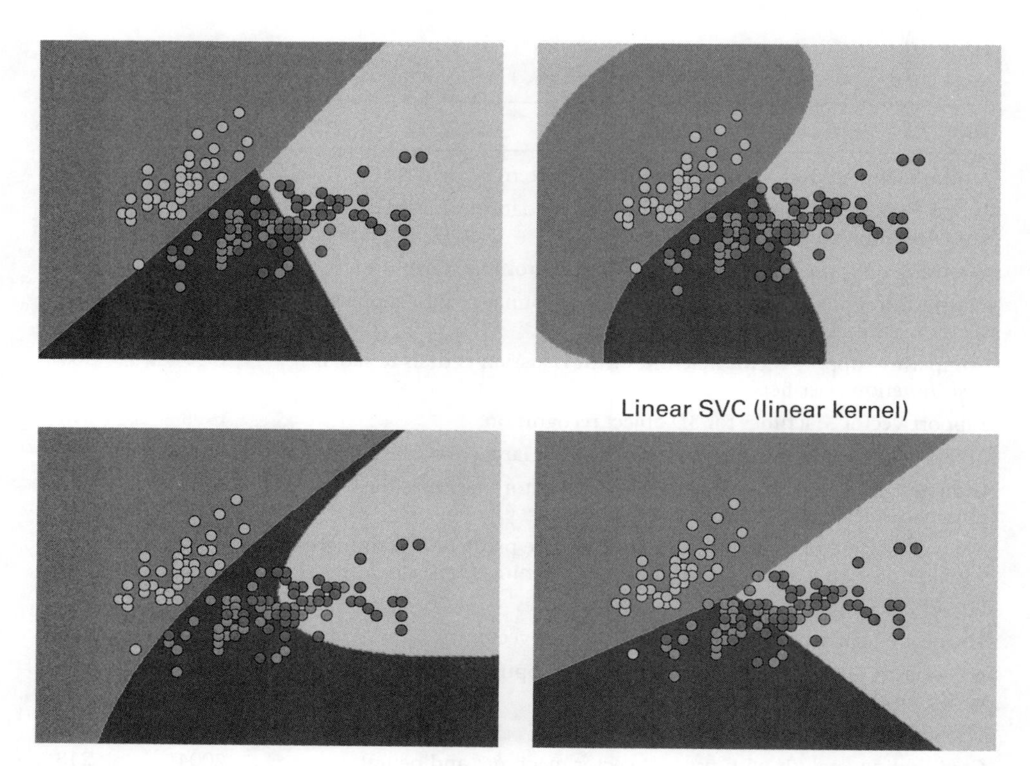

Figure 6.4
Support vector machine on `iris` dataset.

the underlying intuition of the technique is much older and can be found in the models developed by the British statistician R. A. Fisher during the 1930s. Fisher developed the "first pattern recognition algorithm" (Cortes and Vapnik 1995, 273), the "linear discriminator function" (Fisher 1936), to deal with problems of classification and demonstrated its efficacy on the taxonomic problem of discriminating or classifying the irises observed in W.E. Anderson's `iris` dataset (see above).

In his 1936 article in the *Annual Review of Eugenics*, Fisher comments on similar classification work carried out in craniometry and other related settings: so-called "discriminant functions" had been successfully used to distinguish populations. Fisher wrote, "When two or more populations have been measured in several characters, . . . special interest attaches to certain linear functions of measurements by which the populations are best discriminated" (Fisher 1936, 179). The discriminant functions divide the vector space into "regions labeled according to classification" (Hastie, Tibshirani, and

Friedman 2009, 101). "Decision boundaries" (or sometimes "decision surfaces") often appear as straight lines that divide the vector space into regions of constant classification. These longstanding linear discriminant functions were reconstructed during the 1990s in the form of the support vector machine, giving rise to more labile statements about differences, statements whose augmented flexibility can be glimpsed in table 6.2 in the range of things, facts, and beings running through the titles of the papers.

Differences blur?

Decision boundaries change in two ways in support vector machines. They blur and bend, again affecting what counts as pattern. The support vector machine first of all addresses the problem of how to model differences when differences are blurred. An often repeated illustration of how the support vector machine transforms data appears in Cortes and Vapnik's initial publication simply titled "Support Vector Networks" (Cortes and Vapnik 1995). They demonstrate how the support vector machine classifies handwritten digits drawn from a dataset supplied by the U.S. Postal Service (LeCun and Cortes 2012). Like `iris`, the U.S. Postal Service digits and a larger version from the U.S. National Institute of Standard (`mnist`) are standard machine learning datasets. They have been frequently used to measure the performance of competing learning algorithms. In contrast to `iris`, the `mnist` data are high dimensional. Each digit in the dataset is stored as a 16×16 pixel image. Image classification typically treats each pixel as a feature or variable in the input space. So each digit as represented by 16×16 pixels appears in a 256 dimensional vector space. By comparison, `iris` has five dimensions. Unsurprisingly, there are also many more digits in the U.S. Postal Service Database than flowers in `iris`. The `mnist` dataset has around 70,000. Aside from this dimensional growth, the handwritten digits aptly convey the blurring of differences. Many people can easily recognize slight variations in handwritten digits with few errors. This is despite the many variations in handwriting that skew, morph, and distort the ideal graphic forms of numbers.[11]

11 Neural network researchers have heavily used the MNIST dataset. I discuss some of that work in chapter 8. The handwritten MNIST also appear in *Elements of Statistical Learning*, where they are used to compare the generalization error (see previous chapter) of *k* nearest neighbors, convolutional neural network, and a "degree-9 polynomial" support vector machine (Hastie, Tibshirani, and Friedman 2009, 408). What about the handwritten digits attracts so many machine learning techniques? The logistics of the US Postal Service aside (because the `mnist` datasets continue to be used by machine learners well after the problem of scrawl on letters has been sorted), the variations, regularities, and banal everydayness of these digits furnish a referential locus, whose

In their experiments with digit recognition (shown in figure 6.5), Cortes and Vapnik contrast the error rates of decision trees (`CART` and `C4.5`), neural networks, and the support vector machine working at various levels of dimensionality. Support vector machines deal with blurred differences or continuous variations by superimposing two operations: "soft margins" and "kernelization." Nearly all expositions of the support vector machine including (Cortes and Vapnik 1995) highlight the "soft margin" that runs in parallel to the solid decision boundary. The support vector machine develops Fisher's linear discriminant analysis because it searches for a separating hyperplane in the data. While linear discriminant analysis constructs a hyperplane by finding the most likely linear boundary between classes based on all the data, the support vector machine searches for a hyperplane resting only on those cases in the data that lie near the boundary. It introduces the intuition that the best hyperplane differentiating classes will lie near the cases—the *support vectors*—that are most difficult to classify. Hard-to-classify cases become the "support vectors" whose relative proximities tilt the decision surface in various directions. In contradistinction to the $N = \forall X$ proposition (discussed in chapter 5), the machine learner discards much of the data. In contrast to linear discriminant analysis, as Ethem Alpayadin writes, "we do not care about correctly estimating the densities [probability distributions of variables] inside class regions; all we care about is the correct estimation of the *boundaries* between the class regions" (Alpaydin 2010, 210).

Figure 6.6 has appeared in many slight variations in the last two decades. Such figures diagram classes by different point shapes, and the diagrammatic work of the classifier takes the shape of diagonal lines, the solid line marking the decision surface or hyperplane, and the dotted lines marking the soft margins that separate the two classes. In figure 6.6, the dotted lines represent a margin on either side of a hyperplane (the solid line). The support vector machine finds the hyperplane for which that margin or perpendicular distance between the margins is greatest. Of all the slightly different planes that might run between the two classes shown in that figure, the maximum

---

existence as facts, things, or events in the world is less important than the relations of similarity and differences it poses. The field of digits becomes a site of differentiation not only of digits—the machine learners attempt to correctly classify the digits—but of the authority of different machine learning techniques and approaches. They become ways of announcing and delimiting the authority, the knowledge claims, or "truth" associated with the machine. The many uses of the `mnist` data documented by (LeCun and Cortes 2012) suggests something of the ancestral probabilizaton of such datasets.

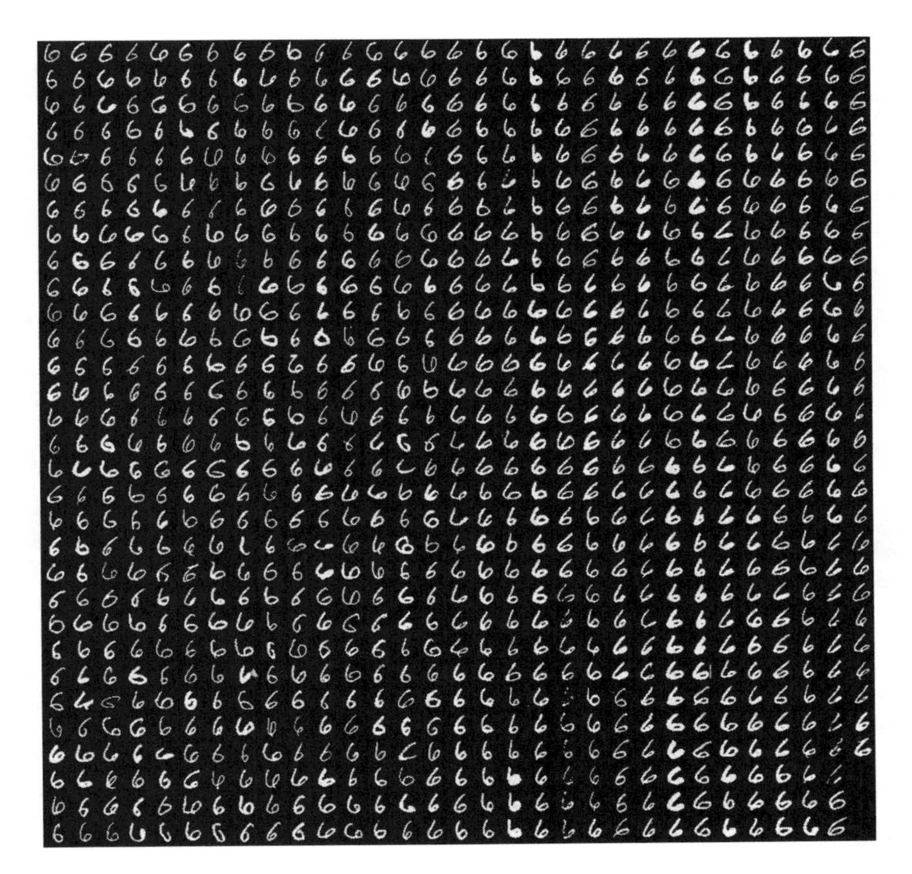

Figure 6.5
MNIST postal digits: sample of "6's."

separating hyperplane lies at the greatest distance from all the points of the different classes. The support vector classifier modifies the idea of the optimal separating hyperplane by accommodating inseparable or overlapping classes. This is something that other machine learners (e.g., the perceptron) cannot do.

Although the geometrical intuition here is that some data points (cases or observations) will lie on the opposite side of the decision surface to where they should be, the distance they lie on the wrong side of the separating hyperplane will be as small as possible. How are the lines showing in figure 6.6 calculated? Locating the optimal separating hyperplane and a limited number of permitted misclassifications presents a complicated optimization problem. As *Elements of Statistical Learning,* following

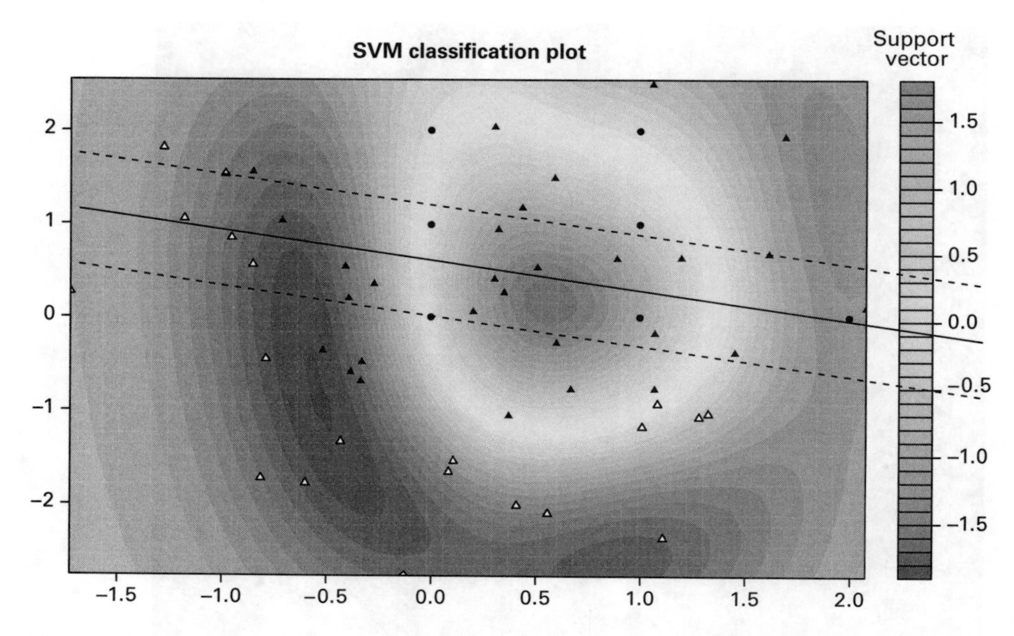

Figure 6.6
Margins in a support vector machine. The margins ignore most of the data and only focus on the hard-to-classify cases.

Cortes and Vapnik's formulations, formalizes it, the problem can be stated in terms of minimization:

$$L_P = \frac{1}{2} \parallel \beta \parallel^2 + C \sum_{i=1}^{N} \xi_i - \sum \alpha \left[y_i \left(x_i^T \beta + \beta_0 \right) - (1 - \xi_i) \right] - \sum_{i=1}^{N} \mu_i \xi_i \tag{6.2}$$

(Hastie, Tibshirani, and Friedman 2009, 420)

In equation 6.2, the optimization problem is to minimize L_P, the Lagrange primal function with respect to β, β_0, and ξ_i (Hastie, Tibshirani, and Friedman 2009, 420). In this complicated optimization problem (one that is difficult to understand without extensive mathematical background), familiar elements include the parameters β in the linear form of $x_i^T \beta + \beta_0$, which is the equation defining the hyperplane, as well as triple operations of addition ($\sum$) of all the values ξ_i, which calculate the distance that each case is on the wrong side of the margin. Correctly classified cases will, therefore, have $x_i = 0$.

As always with mathematical functions, their diagrammatic relations, and the way in which they contain both the generalizing regularities (algebraic icons and indexes, the linear equation, the repeated summation of all values, the shaping parameters) and forms of variation (the presence of the misclassification measures ξ_i), merit our attention. What if elements that lie on the wrong side of the hyperplane were allowed? If that were possible, then the support vector machine could deal with inseparable or overlapping classes, and hence with blurred patterns of difference. Given that a support vector machine permits instances that lie on the wrong side of the separating hyperplane, irregular differences no longer function as errors (as they would appear in most linear classifiers, such as linear discriminant analysis and logistic regression), but as elements in a "soft margin" designed to accommodate inseparability and indistinctness. Equations such as equation 6.2 connect the diagrammatic intuition of a separating hyperplane with a set of second-order observations controlled by parameters such as C, which effectively controls the size of the margin, and α, which effectively bounds the proportion by which a predicted instance can be on the wrong side of the margins that define the hyperplane. In other words, as we have seen previously in cost-function optimization (see chapter 4), the learning in the machine consists of finding a way to transform data into differences according to constraints.

Bending the decision boundary

The support vector machine reinstates a linear decision boundary as the enunciative mode of difference. Yet it transforms that boundary. In the abstract of their 1995 paper, Cortes and Vapnik briefly describe how the support vector machine generalizes the flatness of the linear decision surface:

> The support-vector network is a new learning machine for two-group classification problems. The machine conceptually implements the following idea: input vectors are non-linearly mapped to a very high-dimension feature space. In this feature space a linear decision surface is constructed (Cortes and Vapnik 1995, 273).

Another example of vector space transformation (discussed in chapter 3), this "very high-dimension feature space" is explicitly made to support "a linear decision surface," just as Fisher's linear discriminant analysis had. But this linear decision surface is now located amid a nonlinear mapping of the data.[12] Cortes and Vapnik's support vector machine constructs a new domain—"a very high dimension feature space"—where

12 As we have seen on several occasions, the vector space invites a certain form of classification based on the search for the best line, the line of best fit, or the most discriminating line, the line that best divides things from each other. Linear regression is not called "linear" for no reason. Fisher's "discriminant functions" were later called "linear discriminant analysis" for the same reason: they divide the vector space into different regions ("decision regions") separated by "linear

inseparable differences start to disentangle themselves. The constructed dimensions do not derive from new sources of data. Instead, the support vector machine transforms the vector space into a much higher dimension.

As Vapnik writes in the preface to the second edition of *The Nature of Statistical Learning Theory* (Vapnik 1999, vii), "In contrast to classical methods of statistics where in order to control performance one decreases the dimensionality of a feature space, the SVM dramatically increases dimensionality" (vii). From the standpoint of pattern recognition, this often vastly augmented vector space should make it harder to locate patterns. A linear or planar decision surface in high-dimensional space maps onto a curving even labyrinthine decision boundary when projected back onto the original vector space (see the curving decision boundaries in figure 6.4). In certain cases, machine learners multiply dimensions in data in the name of differentiation, classification, and prediction. Many of the techniques that have accumulated or been gathered into machine learning flatten variations and differences into lines and planes but not always by reducing them. In fact, random forests, neural networks, and support vector machines exemplify a countermovement that maximizes variety in the name of differentiation.[13] Research in machine learning, whether it has been primarily statistical, mathematical, or computational, countenances and addresses problems of nonlinear classification through *dimensional expansion.*

The powerful augmentation characteristic of the support vector machine works through diagrammatic substitution. Consider the expression shown in equation 6.3:

$$L_D = \sum_{i=1}^{N} \alpha_i - \frac{1}{2} \sum_{i=1}^{N} \sum_{i'=1}^{N} \alpha_i \alpha_{i'} y_i y_{i'} \langle h(x_i), h(x_{i'}) \rangle \tag{6.3}$$

(Hastie, Tibshirani, and Friedman 2009, 423)

In equation 6.3, a rewriting of equation 6.2 occurs through the substitution of a product $\langle h(x_i), h(x_i') \rangle$ for x. All of the data x are mapped using some function $h(X)$ into a new higher dimensional space. What would be the value of a more complicated

---

decision boundaries" (Alpaydin 2010, 53). Almost all machine learners are aware of and try to address the idealism or abstraction of the line or plane.

13 Despite the in-principle commitment to any form of function, machine learning strongly prefers forms that can either be visualized on a plane (using the visual grammar of lines, dots, axes, labels, colors, shapes, etc.) or can be computed in form of matrix or vectorized calculations focused on planes. Many of the techniques that grapple with complicated datasets seek to reduce their dimensionality so that lines, planes, and regular curves can be applied to them: multidimensional scaling (MDS), factor analysis, principal component analysis (PCA), or self-organizing maps (SOM) are just a few examples of this.

space? As Leo Breiman writes in his account of the development of the support vector machine:

In two-class data, separability by a hyperplane does not often occur. However, let us increase the dimensionality by adding as additional predictor variables all quadratic monomials in the original predictor variables.... A hyperplane in the original variables plus quadratic monomials in the original variables is a more complex creature. The possibility of separation is greater. If no separation occurs, add cubic monomials as input features. If there are originally 30 predictor variables, then there are about 40,000 features if monomials up to the fourth degree are added (Breiman 2001a, 209).

The extravagant dimensionality realized in the shift from 30 to 40,000 variables vastly expands the number of possible decision surfaces or hyperplanes that might be instituted in the vector space. The support vector machine, however, corrals and manages this massive and sometimes infinite generation of differences at the same time by only allowing this expansion to occur along particular lines marked out by *kernel functions*. Although the support vector machine maintains a commitment to the separating hyperplane, a linear form albeit with soft margins, it reconstitutes that plane in newly created vector spaces constrained by certain key structural features that render them computationally tractable. On the one hand, it promises infinite expansion and associated freedom from the rigidity of lines. On the other hand, the expansion can only countenance a limited range of movements prescribed by the kernel functions (polynomial, radial, etc.).

Elements of Statistical Learning puts it this way:

We can represent the optimization problem and its solution in a special way that only involves the input features via inner products. We do this directly for the transformed feature vectors $h(x_i)$. We then see that for particular choices of h, these inner products can be computed very cheaply (Hastie, Tibshirani, and Friedman 2009, 423).

The terminology here takes us back to the vector space (see chapter 3) that machine learning inhabits. The "inner product" or "the convolution of the dot-product" described by Cortes and Vapnik come from this space, in which the distances or alignments between whatever can be rendered as a vector can be calculated *en masse*.

Top 10 Algorithms for Data Mining (Wu et al. 2008), a widely cited computer science account of data mining, justifies the operation in relation to entangled differences:

The kernel trick is another commonly used technique to solve linearly inseparable problems. This issue is to define an appropriate kernel function based on the *inner product* between the given data, as a nonlinear transformation of data from the input space to a feature space with higher (even infinite) dimension in order to make the problems linearly separable. The underlying justification can be found in *Cover's theorem* on the separability of patterns; that is, a complex pattern classification problem case in a high-dimensional space is *more likely* to be linearly separable than in a low dimensional space (Wu et al. 2008, 42).

The "kernel trick" that overcomes inseparability remaps an already transformed vector space—the inner product of all the vectors in the data—into a higher dimensional space defined by functions such as $f(x) = x_i^2 + x_i^3$. The trick is no simple technical trick because as Cortes and Vapnik point out, it relies on substantial mathematical developments in the 1960s. "The idea of constructing support-vector networks comes from considering general forms of the dot-product in a Hilbert space (Anderson & Bahadur, 1966)," write Cortes and Vapnik (Cortes and Vapnik 1995, 283). It is a trick, however, in the sense that it "can be computed very cheaply."[14]

What does the transformed feature space combined with the computational shortcut of the inner product do in practice? Describing the generalization error—the errors made when a model classifies hitherto unseen data—Cortes and Vapnik highlight the growth in dimensionality introduced by the technique of the support vector in classifying the handwritten numbers of the `mnist` data. They recount how the technique exponentially increases the dimensionality of the feature space and how the error rate on difficult-to-classify handwritten digits drops correspondingly. When the feature space has 256 dimensions (the given dimensions of the 16×16 pixel digits), the error rate is around 12%. As the dimensionality grows to 33,000, a million, a billion, a trillion, and so forth (up to 1×10^{16} dimensions), the error rate drops to just over 4%, close to the errors made by "human performance" (2.5%) (Cortes and Vapnik 1995, 288).

Instituting patterns

The engineered movement of various machine learners do not simply discover differences. They assemble, construct, identify, and optimize distributions or patterns of difference. They do it in different ways. Sometimes they take for granted the possibility of identifying differences in data, as if all differences must be visible and legible given the right partition. At other times, intrinsic inseparability is taken into account as part of the pattern. The power of the support vector machine to do this is limited but constructive. It can deal with various forms of inseparability by taking the difficult-to-classify boundary cases as the basis of a device that brings differences into being. It deals with problems of nonlinearity by increasing the dimensionality of the data and looking for separations in the higher dimensional space.

What do we learn about differences from decision trees and their development into random forests, or from linear discriminant analysis and its reformalization as the

14 This cheapness appeared already in the cat machine learner discussed in the introduction. Heather McArthur's `kittydar` cat image classifier implemented a support vector machine in Javasript that runs in a browser. Cats are classified cheaply there.

support vector machine in some of their operational specificities? First, machine learners multiply and generate patterns. We could also have tracked the movement between the perceptron (Rosenblatt 1958) and the "deep learning" convolutional neural networks (Hinton and Salakhutdinov 2006) of more recent practice (see chapter 8). Second, the ways that machine learners address differences bears witness, I have been suggesting, to the emergence of a new enunciative mode that disperses patterns as the visible form of difference into a less visible but highly operational space. A single decision tree becomes thousands in a random forest. A relatively small number of dimensions in the vector space becomes potentially infinite in the convolutional dot products and kernel functions of support vector machines. Models that sought to encompass or fit everything in the data, including the outliers within a single probability distribution, instead dwell on the difficult-to-classify, erroneous, or borderline instances amid the massive normalized accumulations of event.

What counts as pattern today? The visually interpretable shape of a decision tree cascades into the statistically observable trade-offs between fine-grained classification and cost-complexity considerations, between recursive differentiation and general sparsity. The sharp separating lines and planes that allow linear models to become classifiers in the classic techniques, such as linear discriminant analysis, find themselves displaced into soft margins, or into newly constructed and sometimes inordinately dimensioned feature spaces that can only be traversed by virtue of the kernel functions and their computationally tractable inner products.

What does it matter if pattern disperses into operations? From the standpoint of critical thought, it might be that learning to find dispersed patterns only intensifies a tendency "to see differences in degree where there are differences in kind" (Deleuze 1988a, 21). It would perhaps, however, be redundant to assert the primacy of differences in kind. A more constructive and experimental challenge lies in exploring differences in kind within the computed differences of degree active in machine learning. Despite their quite different ways of partitioning, separating, or propagating differences, support vector machines and decision trees define possibilities of grouping and assembling differences, sometimes through purifying, sometimes through bending and blurring, and sometimes through multiplying them. These groupings and assemblings attract, generate, and accumulate propositions. The grouping and assembling, despite its commonalities, is not homogenous. It does not have the coherence of a science, it uses different systems of formalization (the cross-entropy measures of the decision tree, the tunable soft margins, and the kernel functions of the support vector machine), and it disperses in different ways across knowledge practice (the decision tree with its commercial uptake in data mining vs. the support vector machine's heavy use in image recognition and classification).

7 Regularizing and Materializing Objects

Science is concentrated in an area of knowledge it does not absorb and in a formation which is in itself the object of knowledge and not of science (Deleuze 1988b, 19).

What is the materiality of machine learning? The opening pages of *Elements of Statistical Learning* present four somewhat excessive things or objects—spam email, handwritten digits, prostate cancer, and DNA Expression Microarrays—and list six examples (document classification, image recognition, risk of heart attack, stock price prediction, risk factors for prostate cancer, and glucose estimates for diabetics) (Hastie, Tibshirani, and Friedman 2009, 1–7). What happens to things such as prostate cancer, handwritten digits, or stock prices when machine learners ply them? Although machine learners flourish in the thick of a control crisis (Beniger 1986), I will suggest here that machine learning also occasions viscous multitemporal and interobjectively distributed enactments of things such as financial markets, media platforms, chronic diseases, and living things. These hyperobjects all epistemically, infrastructurally, economically, and socially individuate through machine learning.

The last of the vignettes, the DNA microarray, comes from the life sciences. It attracts a whole-page color figure—a heatmap (Hastie, Tibshirani, and Friedman 2009, 7).[1] DNA, genes, genomics, and proteomics then more or less disappear from view for the next 503 hundred pages of the book (aside from a brief mention in the context of cross-validation), only to abruptly reappear in a discussion of unsupervised machine learning techniques (k-means, agglomerative, and hierarchical clustering; chapter 14, where the DNA microarray data are reanalyzed using hierarchical clustering), and then again, and much more extensively, in a final chapter (chapter 18) new to the second edition of the book on "High Dimensional Problems." Apart from one passage where Hastie and co-authors develop a document classifier for their own journal articles, every example

1 For a discussion of the history of heatmaps and their place in contemporary science, see (Mackenzie 2013a).

in the added chapter 18 comes from genomic science, a scientific field that largely begins to take a recognizable shape in the late 1990s as both sequence data and high-throughput DNA-analysis devices, particularly microarrays, become widely available.

Operational formations often include scientific fields. Alongside the problems of spam email filtering and image or handwriting recognition, scientific research into biological processes constitutes a major reference point and, I will suggest, an axis of materialization for machine learning. In the archaeology of its operational formation, we could say that the scientific domain of genomics has a strongly referential effect on machine learning. What is a referential? For Foucault, a referential "forms the place, the condition, the field of emergence, the authority to differentiate between individuals or objects, states of things and relations that are brought into play by the statement itself; it defines the possibilities of appearance and delimitation of that which gives meaning to the sentence, a value as truth to the proposition" (Foucault 1992 [1966], 91). The referential thereby forms an integral part of the enunciative function, the element that coordinates sites, subject positions, enunciative modalities, forms of accumulation, and differentiation in an operational formation.

Why does the referential of machine learning matter? When hyperobjects are machine learned, they are reconstituted (in vector space, as optimization problems, probability distributions, and patterns of difference). Conversely, as I will suggest in this chapter, they become a site of materialization, cross-validation, and regularization for machine learning in its production of knowledge. But that referential status, which authorizes and imbues statements with value, comes at a cost. The plurality or multiplicity of the hyperobject—genomes, stock prices, and so on—will be regularized and ranked, reused, and transcribed by machine learners over and over to lend coherence to the operational formation and its system of statements.

Genomic referentiality and materiality

`gaagctccac accagccatt acaaccctgc caatctcaag cacctgcctc tacaggtacc` (NCBI 2016).

In contrast to industry, commerce, media, and government, where much that happens is obscured from view, the great virtue or genomic science is the relative openness of its workings and its resolute insistence on DNA as the primary form of data. The fact that data practices are relatively generic and accessible means that critical research into transformations associated with genomic data and knowledge can accompany nearly every aspect of practice.

Genomic data exhibits some specific features. The first feature concerns what I earlier called data strain. Genome data, a tiny fragment of which is shown above, inflates the vector space. Genomics generates new versions of the now familiar problems of data dimensionality. The abundance, diffusion, heterogeneity, or impaction of genomic data thwarts its examination, tabulation, and regulated circulation. Genomics data also present unusual ratios of accumulation and sparsity. Clinical genomics in particular generates datasets that are lavishly furnished with features but often quite meagrely supplied with clinical cases or observations. In the shorthand typical of machine learning terminology, p is larger than N: "the number of features p is much larger than the number of observations N, often written $p \gg N$" (Hastie, Tibshirani, and Friedman 2009, 649). This strains statistical methods that rely on the $\forall X$ "Law of Large Numbers" (Hacking 1990, 99–104), which holds that the accuracy of statistics tends to increase with more observations.

Since the early 19th century, biology and cognate disciplines have sought to explore problems of time, genesis, duration, activity, and process in a broad spectrum of living things. Contemporary genomics seeks to elicit, as many commentators have noted, knowledge of biological, evolutionary, biomedical, and environmental processes from the long DNA sequences comprising genomes. The primary object in genomics is a $\forall X$ data form, the genome, the full complement of DNA in an organism. Genomes vary in size from the 2000 DNA base pairs of a virus, the 3.2 Gb (gigabase pairs) of humans through to the 130 Gb of the lung fish. The founding premise of genomic science is that a dataset comprising the complete sequence of DNA potentially rebases knowledge of many different biological processes, ranging from evolution (phylogeny), development (ontogeny), metabolism, structure, and pathology. If nothing else, genome comparison promises knowledge of the 3.8 billion years of evolution of species differences and population diversity. In all of these respects, since at least the 1980s, DNA sequences have served as the common substrate for many different scientific experiments, technical developments, cyber-infrastructures, and, needless to say, biological imaginaries oriented around the problems of control.[2]

The genomic premise has an ineluctably promissory association with knowledge economy. Prior to whole genome sequencing projects initiated in the 1990s such as the Human Genome Project (HGP), biologists worked only with selected DNA sequences, especially those associated with genes and the proteins that they code. By contrast, the genome, with all its repeated, redundant, and slightly varying patterns of DNA, bears the traces of long evolutionary mixing and constitutes a hypercomplex

2 A large and diverse social science and humanities literature now exists around genomics. I draw on some of that literature as general background here, especially Sunder Rajan (2006), Thacker (2005), Stevens (2011), Leonelli (2014), and Haraway (1997), but largely do not address it directly.

functional process whose exquisite sensitivity to changing conditions—a slight change in light reaching leaf cascades can be traced in patterns of DNA transcription—forms an extreme case for any operational sense of function. The functioning of genomes symbolizes a deeply interconnected relationality in life sciences and becomes the test case for the learning capacities of machine learners.[3]

As referentials, genomes present a mixture of unregulated abundance and seeming homogeneity. DNA sequences exist in great abundance (in databases and, increasingly, from the cheaper and more compact sequencing instruments), yet even determining how DNA sequence fragments should be ordered in a genome—let alone how they make sense as some biological function—is much harder. DNA sequences are assembled as genomes and genomic datasets via statistical models. "Genome assembly continues to be one of the central problems of bioinformatics," write the authors of a recent scientific review of the techniques of constructing whole genomes from DNA sequencer data (Henson, Tischler, and Ning 2012). Even the elementary data form of the genome as DNA base pairs is a highly algorithmic construct. No existing sequencing technology

3 As data forms, genomes have a problematic mode of existence. They resemble cat images on the internet. As a data form, genomes are remarkably homogenous. They are one-dimensional strings of letters corresponding to the well-known four nucleic acids (`g`, `a`, `t`, `c`). Although many earlier tabulations of variation, difference, groups, types, and relations are woven through the life sciences, genomes have for the last several decades mesmerized biological sciences as a way of analyzing and redistributing the confused multiplicities associated with living things. The raw data for genomes come from the sequencing of DNA obtained from various organisms (e.g., viruses, bacteria, plants, fish, animals, and humans). The sequencing of DNA, especially DNA that encodes the proteins that pervade biological processes, that structure tissues or assemble in complicated metabolic pathways, has been the concern first of molecular biology (mainly in the 1970s and 1990s) and more recently genomics (post-1990). In molecular biology, DNA sequences were carefully elicited (using the experimental techniques, for instance, of Sanger sequencing) and then compared with already known sequences of DNA to identify similarities that might have biology significance (e.g., evolution from a common ancestor). In genomics, DNA sequences generally originate from increasingly high-throughput sequencers that output massive datasets (see (Stevens 2013; Mackenzie et al. 2015). Given both the accumulated store of already sequenced DNA and the increasingly viable practices of sequencing all of the DNA in a given organism, genomics has promised a much wider and more detailed understanding of biological complexity than any previous life science had been able to obtain. With genomes in hand, biologists for the first time would be in a position to build models of entire domains of biology, domains that previously could only be explored through painstaking experiments targeting specific cells, molecules, biochemical reactions, and networks. The vast yet somewhat dispersed knowledges of the life sciences might be reordered and aligned on a new extensive yet quite homogenous backbone of the genome read out as billions of DNA base pairs.

produces a genome as a single sequence, as a vector (in the sense described in chapter 3). Instead, sequencing produces random sets of sequence fragments of various lengths that have to be assembled into a complete genome algorithmically.[4]

Between pre- and postgenomic science, the status of significant differences in genomes shifted. Pre-HGP biology understood the significant differences between individual organisms largely in terms of gene alleles responsible for variations in phenotypes. Biological differences, and disease in particular, stemmed from different forms of genes. Understanding disease meant finding the disease genes. Even prominent proponents of genomics, such as Leroy Hood, writing "Biology and Medicine in the Twenty-First Century" in 1991, envisaged genomics as a way of simplifying "the task of finding disease genes" (Hood and Kevles 1992, 138). Across the life sciences, genes were the object of annotation, labeling, and description. Two decades after the inception of whole genome sequencing, genomes present a different image of variation. According to Nikolas Rose, writing more recently, "There is no normal human genome; variation is the norm" (Rose 2009, 75). "In this new configuration," he writes, "what is

4 Whole genome assembly as reported for the initial draft of the human genome in 2001 (Venter et al. 2001; Lander et al. 2001) or for the model biological organism, *Drosophila* (Myers et al. 2000), was not at the time understood as a machine learning problem. The task of whole genome assembly from DNA fragments was seen as probabilistic in the sense that the aim is to assemble the often millions of short sequence fragments in an order that is most likely to occur. Even prior to the first full human genome assembly, genomic science had made heavy use of probabilistic models in aligning DNA (and protein amino acid) sequences. Richard Durbin, Sean Eddy, Anders Krogh, and Graeme Mitchison's highly cited *Biological Sequence Analysis: Probabilistic Models of Proteins and Nucleic Acids* (Durbin et al. 1998) was based almost entirely on Hidden Markov Models, a way of modeling a sequence of states that *Elements of Statistical Learning* treats at chapter length (see (Hastie, Tibshirani, and Friedman 2009, chapter 17)). Although sequence alignment was regarded as a deeply algorithmic and statistical problem in the former volume, it is not at all formulated in the language of machine learning. There is little discussion of cost functions, vector spaces, optimization, problems of generalization, and supervised or unsupervised learning. David Haussler, a key bioinformatician in the first draft of the human genome, in his work explicitly sought to bring machine learning methods to bear on biology and continues to do so. See Zerbino, Paten, and Haussler (2012) for a review of the relevance of machine learning to genomic science. The practical problem here is that genomes contain swathes of duplicated regions that make assembling sequences in good order a severe challenge. Although sequence alignment algorithms have long used algorithmic approaches (known as dynamic programming) to score the similarity between two given DNA sequences, assembling the millions of DNA sequences produced by contemporary sequencers has necessitated entirely new techniques (shifting, e.g., from the overlap-layout-consensus model to the de Bruijn graph-based path models (Pevzner, Tang, and Waterman 2001).

required is not a binary judgment of normality and pathology, but a constant modulation of the relations between biology and forms of life, in the light of genomic knowledge." The emphasis in Rose's formulation falls on constant modulation of the relations between biology and forms of life. If postgenomic science departs from the understanding that there is no single genome but many genomes, then according to Rose, variation becomes of primary interest. Pursuit of variation remakes the genome into "a form whose only object is the inseparability of distinct variations" (Deleuze and Guattari 1994, 21).

Whatever knowledge subsequently derives from a genome (genes, mutations, evolutionary relationships, variations associated with disease, heredity, or individual identity), genomic data and hence the genome itself as a scientific hyperobject is deeply probabilistic. From DNA sequence assembly through the ancestral probabilization embodied in heavily used biological databases, to the entangled reliances on accumulated biological knowledges, genomes present a particularly challenging site of machine learning activity.

Elements of Statistical Learning's invocation of DNA-related data, therefore, is no arbitrarily chosen example amid the general proliferation of settings, domains, cases, and examples typically found in machine learning pedagogy. In multiple dimensions and directions, genomics—the scientific project of operating on the whole DNA complement of organisms—is a tightly coupled referential for machine learning even if relatively few machine learners have, to date, managed to work with whole genome sequence data. The relatively long-established referential entanglement (at least 25 years and perhaps more) of genomics and machine learning is strategically important in the generalization of machine learning, in the processes whereby techniques, with their specific forms of articulation, statement, and making-visible, propagate into multiple, once-disparate settings.[5] Like social media platforms or retail spaces with their many visitors, genomes, I would suggest, provoke a multiplicity of machine learners to bind to them like antibodies to an antigen (or an allergen). Genomes function as regularizing hyperobjects for machine learning.

5 Signal processing is another such domain. Many of the techniques now prominent in machine learning developed in parallel with signal processing, where the encoding and decoding of signals has long been seen as a problem of pattern recognition amenable to statistical calculation. In some specific cases, such as Hidden Markov Models, the same techniques seem to appear almost simultaneously in disparate domains. Hidden Markov Models appear in genomics (as part of the problem of sequence alignment) at the same time as they begin to appear in digital signal processing for wireless communication, video image compression (Mackenzie 2010), and, above all, in speech recognition (Rabiner 1989).

Table 7.1
The top 20 disciplines of the top 5000 cited research publications in machine learning, 1990–2015

	Frequency	Discipline
1	1302	computer science, artificial intelligence
2	924	engineering, electrical & electronic
3	520	statistics & probability
4	401	computer science, information systems
5	344	computer science, interdisciplinary applications
6	332	biochemistry & molecular biology
7	282	mathematical & computational biology
8	259	biotechnology & applied microbiology
9	258	biochemical research methods
10	228	neurosciences
11	218	computer science, theory & methods
12	199	radiology, nuclear medicine & medical imaging
13	198	multidisciplinary sciences
14	188	genetics & heredity
15	181	immunology
16	157	ecology
17	155	imaging science & photographic technology
18	125	automation & control systems
19	122	engineering, biomedical
20	106	computer science, hardware & architecture

The genome as threshold object

From 1990 to 2015, biology, and particularly molecular and then genomic biology, had high visibility in the machine learning research literature (see table 7.1). After the leading machine learning disciplines (computer science, electronic engineering, and statistics), molecular biology, genomics, and bioinformatics attract most academic journal citations and publications associated with machine learners. Half of the most cited research literature has a biomedical or life science referentiality. This may be because genomes and human disease in particular are premiere scientific hyperobjects like the human brain, dark matter, global climate, or fundamental particles in contemporary sciences. But it might also be the case—and I will pursue this line of argument here—that genomes, with all their operational and functional complexity, come into play, are potentialized and regulated, and take on promissory epistemic value as zones of collective individuation in machine learning. In terms of contemporary biological knowledge production, the transformation of biology into a data-intensive science (Hey, Tansley, and Tolle 2009; McNally et al. 2012) is tightly entangled with machine learning in processes of cross-validation.

In the generalization of machine learning, the genomic referential marks a threshold of materialization. The archaeological approach to materiality is somewhat unusual. Given his interest in operational statements, Foucault understands materiality as a regulatory process operating in an enunciative function. Foucault writes:

> The rule of materiality that statements necessarily obey is therefore of the order of the institution rather than of the spatio-temporal localization; it defines possibilities of reinscription and transcription (but also thresholds and limits), rather than limited and perishable individualities (Foucault 1972, 103).

In the archaeology of an operational formation, locating specific practices, places, and times of reinscription, transcription, and possibilities of reuse carries more weight than any direct conceptual account of materiality. What would materiality in this sense mean for machine learning?

Genomes are both a challenge to the capacity of machine learning to produce scientific knowledge (as distinct from, say, the unstable commercial knowledge of a credit risk model) and a cross-validation of machine learning as a relevant knowledge practice. Genomes first of all authorize infrastructural vectorizations such as computational clusters, grids, arrays, and clouds. For instance, the Google Compute Engine, a globally addressable ensemble of computers typical of recent distributed commercial computing architectures, was briefly turned over to exploration of cancer genomics during 2012 and publicly demonstrated at the annual Google I/O conference. Midway through the demonstration, in which a human genome is visualized as a ring in "Circos" form (see figure 7.1; Krzywinski et al. 2009), the speaker, Urs Hölzle, Senior Vice President of Infrastructure at Google, "then went even further and scaled the application to run on 600,000 cores across Google's global data centers" (Google Inc. 2012). The audience clapped as the annular diagram of a human genome was decorated with a rapidly increasing number of cross-links, accompanied by a snapping sound as it appeared. The world's "3rd largest supercomputer," as it was called by *TechCrunch*, a prominent technology blog, "learns associations between genomic features" (Anthony 2012). Note the language of machine learning: it "learns ... associations between features." We are in the midst of many such demonstrations of "scaling applications" of data in the pursuit of associations between "features."[6]

6 A second significant and equally prestigious example of this infrastructural rescaling might be IBM Corporation's "cognitive computing platform," Watson. Watson, a distributed computing platform centered on machine learning, is hard to delineate or easily describe because it exists in a seemingly highly variable form. Its uses in genomics, pharmaceutical discovery, clinical trials, and cooking are heavily promoted by IBM (IBM 2014). Another would be Amazon Web Services, various cloud computing services, some of which have been heavily used by genomic scientists.

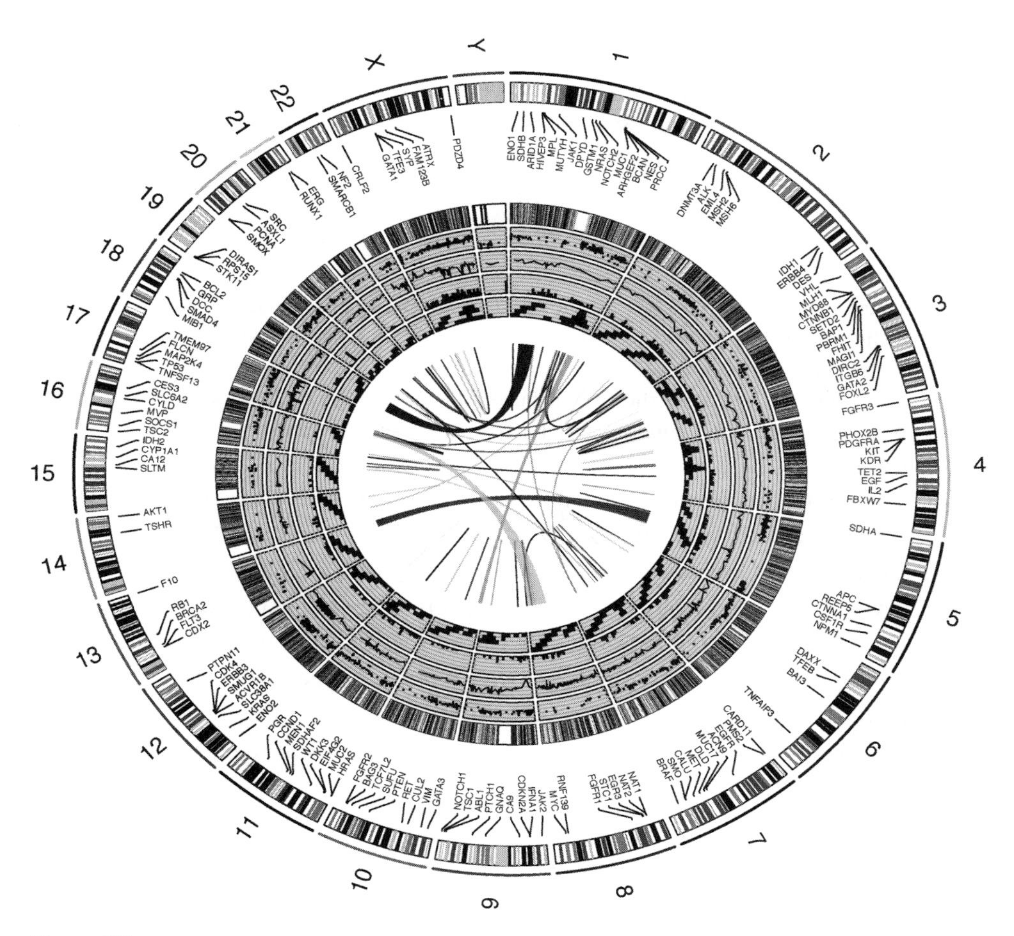

Figure 7.1
A human genome diagrammed using the Circos form. The many tracks of this diagram support a range of graphic forms including scatterplots, heatmaps and histograms all anchored to the ideogram of the 23 chromosomes of the human genome.

The I/O conference audience, largely comprising software developers, could hardly be expected to have a detailed interest in cancer genomics. Their interest was steered toward the immediate availability of computing power: from 10,000 to 600,000 cores in a few seconds. Such drastic infrastructural rescaling attests to the provocation of the genomic referential. The Google Compute demonstration is, I would suggest, typical of how genomes, genes, proteins, and biological sciences more generally authorize differentiation of individuals, events, and things through machine learning. This differentiation is only hinted at in the Google I/O keynote address in Hölzle's talk of genomic features, gene expression, and patient attributes.

The only concrete indication of how what was happening in the demonstration related to machine learning was one mention of the `RF-ACE` (Random Forest-Artificial Contrasts with Ensembles) algorithm. Google's press release emphasizes the distribution of learning across an infrastructure:

> The primary computation that Google Compute Engine cluster performs is the `RF-ACE` code, a sophisticated machine learning algorithm which learns associations between genomic features, using an input matrix provided by ISB (Institute for Systems Biology). When running on the 10,000 cores on Google Compute Engine, a single set of association can be computed in seconds rather than ten minutes, the time it takes when running on ISB's own cluster of nearly 1000 cores. The entire computation can be completed in an hour, as opposed to 15 hours (Google Inc. 2012).

Google repurposes an algorithm developed by engineers at Intel Corporation and Amazon and draws on genomic datasets provided by the Institute of Systems Biology, Seattle, a doyen of big-data genomics. The demonstration animates the `RF-ACE` (a further development of Breiman's random forests discussed in chapter 6) by redrawing a diagram of the genome and redraws it increasingly rapidly as the demonstration scales up to 10,000 cores (or CPUs). A diagram that normally appears statically onscreen or on the printed page of a scientific publication is now animated by an algorithmic process. This confluence of commerce (Amazon), industry (Intel), media (Google), and genomic science (ISB) exemplifies the reinscriptive or transcriptive materiality of machine learning.

Genomic knowledges and their datasets

In the infrastructural materiality of these demonstrations and examples, whether they come from *Elements of Statistical Learning* or from Google Compute Engine, the object of knowledge—genomes, genes, proteins—does not figure in terms of its original discipline or scientific field (typically cancer biology). The scaled-up demonstration of `RF-ACE` on Google Compute assembles only a general system of references among cancer patients, vectorized infrastructures, and predictive classifications. Similarly, the treatment of DNA microarray data in the slightly earlier examples found in *Elements of Statistical Learning* does not principally concern cancer biology as such, but much more the way a group of elements are assembled so as to permit the production of propositions that cross the threshold of scientificity. They may just as well cross different thresholds of knowledge in governmental, market-focused, organizational, or managerial operations.[7]

The plurality of applications can sometimes make it seem that machine learning arrives at the borders of different domains and then proceeds to colonize local knowledges practices. The rule of materiality here would seem to be an epistemic *terra nullius* appropriation, in which existing knowledge forms are rapidly extinguished by machine learners. We have seen previously that ancestral communities of probabilization orient the generalization of machine learning (see chapter 5). Research literature published on machine learning since the early 1990s clusters around problems of plethoric excess—image recognition, document classification, market behavior (as in working out what advertisement to show, whether someone is likely to buy a particular product, etc.). These problems position machine learning amid regimes of communication, the production of economic value, and the regularities of statements (or put in more Foucauldian terms, amid life, labor, and language; see Foucault 1992 [1966]). Where, amid these major regularities, does genomics (arguably the successor of molecular biology) fit? Almost all of the major machine learners, albeit supervised or unsupervised, discriminative or generative, parametric or nonparametric, substantial research activity during the last two or so decades, cross-validate their statements with genomics.

The advent of "wide, dirty, and mixed" data

We can see this referential cross-validation at work in the shape of genomic data. The DNA microarray data extensively modeled in the final chapter of *Elements of Statistical Learning* highlights some elementary problems of shape associated with genomic data. The `iris` dataset (Fisher 1936), perhaps the most heavily used pedagogical dataset in the literature, does not provoke the infrastructural contortions associated with Google

7 In their account of the surprisingly slow shift of microarrays toward clinical practice, Paul Keating and Alberto Cambrosio identify statistics as a kind of bottleneck:

> The handling and processing of the massive data generated by microarrays has made bioinformatics a must, but has not exempted the domain from becoming answerable to statistical requirements. The centrality of statistical analysis emerged diachronically, as the field moved into the clinical domain, and is re-specified synchronically depending on the kind of experiments one carries out (Keating and Cambrosio 2012, 49).

What Keating and Cambrosio describe as "becoming answer to statistical requirements" I would suggest also entails a transformation of statistical requirements in a new operational diagram that reduces some of the frictions associated with existing statistical practice. This operational diagram is machine learning.

Table 7.2
First five rows of Fisher's "iris" dataset

Sepal.Length	Sepal.Width	Petal.Length	Petal.Width	Species
5.10	3.50	1.40	0.20	setosa
4.90	3.00	1.40	0.20	setosa
4.70	3.20	1.30	0.20	setosa
4.60	3.10	1.50	0.20	setosa
5.00	3.60	1.40	0.20	setosa

Compute or, for that matter, the highly sophisticated and quite subtle treatment of gene expression we find in genomics-related machine learning.

Machine learners, in working with `iris`, use the variables from the first four columns shown in table 7.2 to infer the value of the `Species` response variable (as seen in chapter 5, where a decision tree was constructed using this same dataset). The measurements of petals and sepals of the irises of the Gaspé Peninsula in Novia Scotia and their classification into different species is perhaps a typical mid-20th-century biological procedure. Even in the excerpt shown in table 7.2, we can see that the table is quite narrow because it has only a few columns, the data are nearly all of one type (measurements of lengths and widths), and the data are clean (there are no missing values). `Iris` is typical of classic statistics and much biological data prior to genomics in its relatively homogeneity and distinct partitioning.

If `iris` is the conventional statistical form, then how does a genomic dataset differ? One clue comes from descriptions of the `RF-ACE` algorithm, first published in 2009. RF–ACE attempts to deal with "modern data sets" that are "wide, dirty, mixed with both numerical and categorical predictors, and may contain interactive effects that require complex models" (Tuv et al. 2009, 1341). Such algorithms and the "wide, dirty, mixed" datasets they work on have an irregular texture, which, I would suggest, we should try to grasp if we want to understand how genomic data constitute a complex volume "in which heterogeneous regions are superposed" (Foucault 1972, 128). Clues to the irregularity of genomic data come from the various treatments of DNA microarray data in *Elements of Statistical Learning*.

Hastie and co-authors introduce one microarray dataset they use in this way:

> The data in our next example form a matrix of 2308 genes (columns) and 63 samples (rows), from a set of microarray experiments. Each expression value is a log-ratio log(R/G). R is the amount of gene-specific RNA in the target sample that hybridizes to a particular (gene-specific) spot on the microarray, and G is the corresponding amount of RNA from a reference sample. The samples arose from small, round blue-cell tumors (SRBCT) found in children, and are classified into

four major types: BL (Burkitt lymphoma), EWS (Ewing's sarcoma), NB (neuroblastoma), and RMS (rhabdomyosarcoma). There is an additional test data set of 20 observations. We will not go into the scientific background here (Hastie, Tibshirani, and Friedman 2009, 651).

Note that although the number of samples (~80) in the small round blue-cell tumors (`SRBT`) (Khan et al. 2001) dataset is less than the number of flowers measured in `iris`, the number of variables presented by the columns in the table (2308) is much greater. Hastie and co-authors, like the Google I/O demonstration, do not "go into the scientific background." Scientific knowledge per se is not the central concern in machine learning. Rather, genomic data as a field or emergence and differentiation in the production of statements matter. The original publication of this dataset in 2001 (Khan et al. 2001) also made use of machine learning techniques (neural networks, a major topic in chapter 8) precisely to address the diagnostic problem of distinguishing different tumor types without resorting to new experiments or biological knowledge.[8]

The sample of the `SRBCT` data shown in table 7.3 does not readily accommodate the width of the dataset. Unlike `iris`, the thousands of variables simply cannot be displayed on a page or screen. *Wide* datasets are quite common in machine learning settings generally but are particularly common in genomics, where in a given study there might only be a relatively small number of biological samples but a huge amount of sequencer or microarray data for each sample. Many genomic data share this generic feature of width.[9]

8 Khan and co-authors write:

> Gene-expression profiling using cDNA microarrays permits a simultaneous analysis of multiple markers, and has been used to categorize cancers into subgroups 5–8 . However, despite the many statistical techniques to analyze gene-expression data, none so far has been rigorously tested for their ability to accurately distinguish cancers belonging to several diagnostic categories (Khan et al. 2001, 673).

9 Importantly, as discussed in chapter 2 (in terms of the diagonalization running between different elements of code, data, mathematical functions, and indexical signs) and in chapter 3 (in terms of the auratic power of datasets), the fact that these datasets can be so readily loaded and accessed via bioinformatic infrastructures using code written in `R` or `Python` is also a notable feature of their advent in the machine learning literature. Even a social science researcher can quickly write programs to retrieve these data. It attests to several decades, if not longer, work on databases, web and network infrastructures, and analytical software, all, almost without exception, driven by the desire for aggregation, integration, archiving, and annotation of sequence data that first became highly visible in the Human Genome Project of the 1990s. The brevity of these lines of code that retrieve and load datasets—half a dozen statements in `R`, no more—suggests we are dealing with a high-sedimented set of practices, not something that has to be laboriously articulated, configured, or artificed. Code brevity almost always signposts highly trafficked routes in contemporary network cultures. Without describing in any great detail the topography of databases,

Table 7.3

Small round blue-cell tumor data sample (Khan, 2001)

GENE1	GENE2	GENE3	GENE4	GENE5	GENE6	GENE7	GENE8	GENE9	GENE10	GENE11	GENE12	GENE13	GENE14	GENE15
0.77	−2.44	−0.48	−2.72	−1.22	0.83	1.34	0.06	0.13	0.57	1.50	0.39	1.63	0.82	0.01
−0.08	−2.42	0.41	−2.83	−0.63	0.05	1.43	−0.12	0.46	0.16	1.15	0.38	1.56	0.01	0.16
−0.08	−1.65	−0.24	−2.88	−0.89	−0.03	1.16	0.02	0.19	0.50	1.39	−0.53	1.61	−0.21	0.08
0.97	−2.38	0.63	−1.74	−0.85	0.95	1.09	0.82	−0.28	0.99	1.01	0.08	1.05	0.97	−0.17
0.08	−1.73	0.85	0.27	−1.84	0.33	1.25	0.77	0.03	0.28	1.11	−0.39	1.19	0.42	−0.39

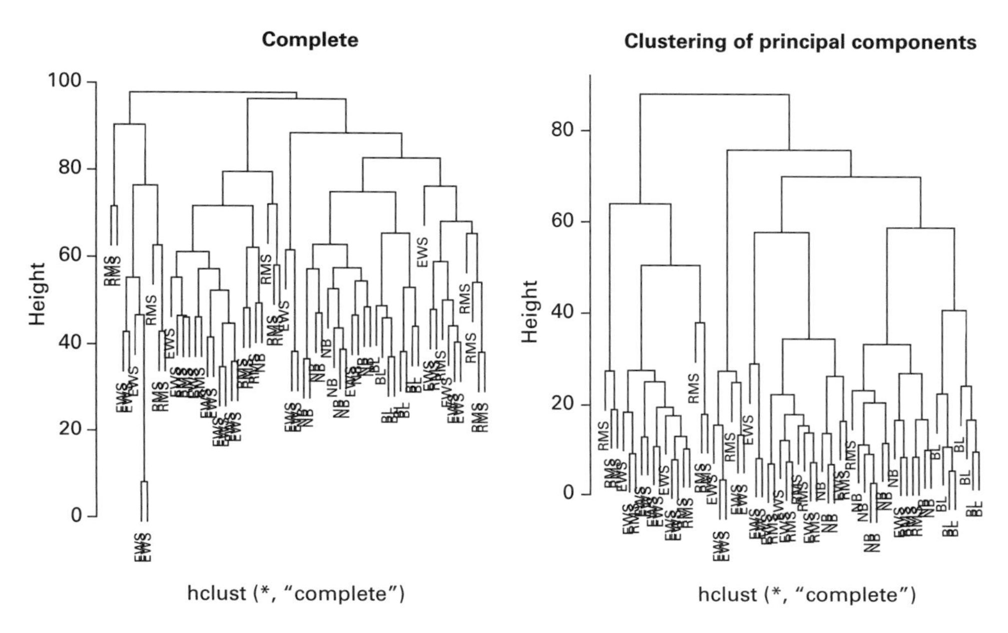

Figure 7.2
Hierarchical clustering of the SBRCT gene expression data.

In contrast to the direct measurements of petals and sepals in the `iris` data, the `SRBCT` data incorporates and diagonally connects many levels of practice. The columns in table 7.3 refer to genes whose levels of expression in different samples are measured and then grouped by comparison to their levels in a reference sample (see the hierarchical clustering of the data shown in figure 7.2). Even the identification of the several thousand genes whose levels of expression are measured by the microarray experiments presupposes much preceding work on DNA sequences and the identification of

---

protocols, and standards woven by and weaving through bioinformatics, the ready invocation of genomic datasets suggests that the mixed, dirty, wide datasets fed to algorithms such as `RF-ACE` or analyzed in (Hastie, Tibshirani, and Friedman 2009) derives from the layered couplings and interweaving of scientific publications and scientific databases developed by biological science over the last three decades. As the code shows, sequence and other genomic data are available to scientists not only as users searching for something in particular and retrieving specific data, but to scientists as programmers developing ways of connecting up, gathering, and integrating many different data points to produce the wide (many-columned), mixed (different types of data), and dirty (missing data, data that are "noisy") datasets, datasets whose heterogeneous and often awkward topography then elicits and invites algorithmic treatment.

protein-coding DNA regions amid the highly repetitive vector of the genome sequence. Highly leveraged knowledge infrastructures underpin such datasets. Considered more diagrammatically, genomes in many ways become less linear or flat than the bare base DNA sequences might suggest.

Cross-validating machine learning in genomics

The linear sequences of DNA data mix and diffuse partly through the archival accumulation that allows them to be superimposed, annotated, and layered, but also through the many efforts to traverse their expanded volume using classifiers and predictive models. A recent review in the journal *Genomics* highlights the increasing bearing of machine learning techniques on genomic science:

> High-throughput genomic technologies, including gene expression microarray, single Nucleotideide polymorphism (SNP) array, microRNA array, RNA-seq, ChIP-seq, and whole genome sequencing, are powerful tools that have dramatically changed the landscape of biological research. At the same time, large-scale genomic data present significant challenges for statistical and bioinformatic data analysis as the high dimensionality of genomic features makes the classical regression framework no longer feasible. As well, the highly correlated structure of genomic data violates the independent assumption required by standard statistical models (Chen and Ishwaran 2012, 323).

Commentary on the "highly correlated structure," not just the volume, of genomic data, points to another referential operation concerning the differentiation of things. Many such statements highlight the incompatibility between a surging multiplicity of data forms and the constraints of existing statistical modeling techniques ("standard statistical models"). So, for instance, Chen and Ishwaran recommend the use of the random forest (RF) because it:

> is highly data adaptive, applies to "large p, small n" problems, and is able to account for correlation as well as interactions among features. This makes RF particularly appealing for high-dimensional genomic data analysis, . . . including prediction and classification, variable selection, pathway analysis, genetic association and epistasis detection, and unsupervised learning (Chen and Ishwaran 2012, 323).

Familiar machine learning vectorization keywords such as "large p, small n" and "high dimensional" pepper their recommendations. But the key terms on the genomics side of this formulation would perhaps be "pathway analysis," "genetic association," and "epistasis." Such biological terms point to forms of relationality associated with biologically interesting processes. Epistasis, for instance, broadly refers to linked gene action, a process that has been difficult to study before high-throughput methods of functional genomics brought shifting patterns of gene expression to light. In contemporary genomic science, these biological processes are increasingly understood in terms of eliciting and modeling the relations between *features* of genomic datasets to classify and predict biological outcomes.

How does machine learning differ from the statistical practice that has underpinned much of modern biology? The analysis of `SRBCT` gene expression in *Elements of Statistical Learning* is symptomatic of a mutual articulation, a cross-validation that entangles genomics and machine learning. The overt arrival of machine learning techniques in genomic research was initially largely concerned with the problem of variations in gene expression (and in fact nearly all of the analysis of genomic data in *Elements of Statistical Learning* explicitly deals with various cases of gene expression). On the one hand, genomics data promises legibility of all the genes in a given organism (~20,000 for humans). On the other hand, the pattern of activity of these genes in time, or any particular point in the life of an organism, cannot be read from the genome but only in time-varying expression, the changes in state, and the variations in closely similar genomes.

Compared with the refined algorithmic craft of whole genome assembly (Venter et al. 2001; Myers et al. 2000; Pevzner, Tang, and Waterman 2001), the handling of the problem of gene expression in machine learning settings can seem rather crudely lacking in biological specificity. Hastie and co-authors almost deprecate scientific knowledge: "We will not go into the scientific background here." Like the authors of the original scientific study (Khan et al. 2001), *Elements of Statistical Learning* treats gene expression profiling largely as a problem of learning to classify differences in disease or other health-related conditions. The many gene expression studies seek to discriminate among different conditions, diseases, or pathologies on the basis of differing levels of gene expression. For machine learners, each gene is a variable whose levels of expression in a given sample may help identify what type that sample belongs to. In the case of the `SRBCT` data, the types include lymphomas, sarcomas, and neuroblastomas.

Like Chen, *Elements of Statistical Learning* begins by addressing the problem of the shape of the data. "Since $p \gg N$," write Hastie and co-authors, "we cannot fit a full linear discriminant analysis (LDA) to the data; some sort of regularization is needed" (Hastie, Tibshirani, and Friedman 2009, 651). What is this regularization? Like the redistribution of classification into a randomized population of machine learners (see chapter 5), regularization governs a potentially unruly plurality through a form of training and observation. Michel Foucault describes the advent of disciplinary power partly in terms of enclosure or individualizing observation but also in terms of techniques of supervising, examining, and, above all, *regularizing* conduct. He writes:

Shift the object and change the scale. Define new tactics in order to reach a target that is now more subtle but also more widely spread in the social body. Find new techniques for adjusting punishment to it and for adapting its effects. Lay down new principles for regularizing, refining, universalizing the art of punishing (Foucault 1977, 89).

Foucault's description of regularization as a technique of disciplinary power—the formation that emerged in the late 18th century as a way of ordering "massive or transient pluralities" (143) in Western European societies—seems a long way from microarray gene expression data. Yet the data in genomic and other referentials (transactions, social media, etc.) display some of the traits—massiveness, transience, plurality—that Foucault identifies as key targets of regulation for the operations of disciplinary power focused on the social body or populations. The "target," a term often used in machine learning to describe the type, group, or response being modeled, in genomics is often subtle variations (in gene expression, phylogeny, pathogenesis), and these variations are widely dispersed in genomic sequence data and in the populations it stems from. Foucault's account of supervision (*surveiller*) and penalization as disciplinary techniques responding to "popular illegality" (Foucault 1977, 130) dwells on the capillary network of observations, examining, ranking, test, and gradation that adapt to the surging multiplicities by ordering them in tables. Although the tables of data (see table 7.3 in the microarray gene expression datasets suggest the persistence of the same technique of ordering multiplicities through partitioned observations, the *cells* no longer target individuals under observation but focus on the attributes of a multiplicity in movement, the human genome, for instance, in its many functional states.

"Shift the object and change the scale," Foucault writes, in describing how partitions, segmentations, forms of enclosure, and, above all, ranked classifications target a more subtly distributed nexus of relations in disciplinary power. Often understood in terms of enclosure and surveillance, disciplinary power, according to Foucault, operates through ranking: "discipline is an art of rank, a technique for the transformation of arrangements" (Foucault 1977, 145). "Regularize in a way that automatically drops out features that are not contributing to the class predictions," Hastie and co-authors write (Hastie, Tibshirani, and Friedman 2009, 652) in describing how regularization deals with the problem of too many variables in the microarray datasets. In the many different techniques that *Elements of Statistical Learning* brings to bear on the problem of gene expression—diagonal linear discriminant analysis, nearest shrunken centroids, linear classifiers with quadratic regularization, regularized discriminant analysis, regularized multinomial logistic regression, support vector classifier—essentially the same ordering movement occurs. Regularization rescales the excessive potential relations of the hyperobject—the patterns of expression of genes associated with different types of tumors—by shrinking or dropping the weights or parameters of each gene in the model and examining the effect on the predictions that result. The coefficients or weights of parameters in the model, the β_p values, are ranked by importance and then are either reduced (L_2 regularization) or eliminated (L_1 regularization) if they contribute little to the predictive accuracy of the machine learner. Learning here takes the form of regularization.

A technique called "lasso regression" displays features that might help us grasp how machine learners regularize genomic data. Remember that the linear regression model with its diagonal line or plane running through vector space provides the underlying intuition for many machine learners. We have seen the function in equation 7.1 several times already in different variations, including logistic regression used for classification of types or groupings.

$$\hat{Y} = \hat{\beta}_0 + \sum_{j=1}^{p} X_j \hat{\beta}_j \tag{7.1}$$

In gene expression models, the values of β shown in equation 7.1 map onto the different levels of expression of the many genes indexed by the p columns of the microarray dataset. The model tests how different patterns of gene expression associate with different tumor types. As we have already seen, the number of combinations of genes associated with different tumor types vastly outweighs the number of samples.

The regularized version of the linear regression framework known as "lasso"—Least Absolute Shrinkage and Selection Operator—introduces a different form of training and observation of model construction. This train hinges on the lasso penalty shown in equation 7.2[10]:

$$\widehat{\beta}^{lasso} = argmin_{\beta} \left\{ \frac{1}{2} \sum_{i}^{N} (y_i - \beta_0 - \sum_{j=1}^{p} x_{ij}\beta_j)^2 + \lambda \sum_{j=1}^{p} |\beta_j| \right\}. \tag{7.2}$$

(Hastie, Tibshirani, and Friedman 2009, 68)

10 The original publication of the lasso technique in a paper titled "Shrinkage and Selection via the Lasso" (Tibshirani 1996) has been heavily cited in subsequent literature. Google Scholar counts around 13,000 citations. For a paper published in the *Journal of the Royal Statistical Society*, this is surprisingly high, but attests, I would suggest, to the intense interest in renovating linear models for new problems such as image recognition or tumor classification. Somewhat surprisingly, given its heavy usage in other scientists, Andrew Ng's CS229 machine learning course at Stanford University doesn't mention the lasso.

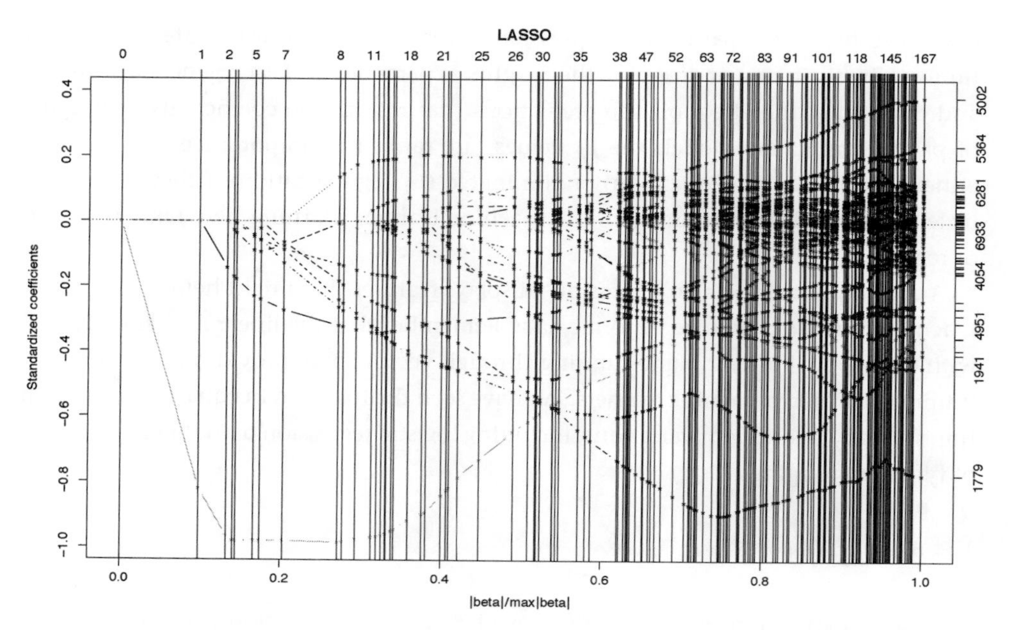

Figure 7.3
Shrinkage path of coefficients for Lasso regression on (Golub, 1999) leukemia data.

Equation 7.2 is notable for the way that it subjects the familiar "residual sum of squares" way of calculating the coefficients to the "penalty" carried by the last part of the equation, $\sum_i^p |\beta_j|$. As Hastie and co-authors write, "the lasso does a kind of continuous subset [feature] selection" (Hastie, Tibshirani, and Friedman 2009, 69). As always, $argmin_\beta$ suggests that the algorithm should optimize the set of values for β that minimize the overall value of the function. It balances the costs of reducing the sum of residual errors shown in the first half of the equation and minimizing the sum of the absolute values of the model parameters β_j in the second part of the function. The optimizing double movement reshapes its expression of the data along a diagonal line drawn as the algorithm gradually introduces and scales all of the features in the vector space **X**, only allowing those variables or features to remain in the set that helps minimize the difference between the predicted response and the actual response. (Figure 7.3 makes something of this scaling diagrammatically visible. In this diagram, the various diagonal lines show how values of coefficients grow and sometimes diminish as the `lasso` process runs. Vertical lines show steps as new variables are included in the model with different values of the control parameter λ.)

Regularization sometimes radically changes the object. In figure 7.3, these changes become a matter of diagrammatic observation. Comparing eight different methods for analyzing the microarray cancer data from (Ramaswamy et al. 2001), Hastie reports that "lasso regression (one versus all)" selects 1,429 of the 16,063 genes in the dataset. The shifted-rescaled object, a set of 1,400 genes, or in the case of the "elastic-net penalized multinomial" model that uses only 384 genes, highlights a drastically reduced subset of the original object. A regularized genome of 384 genes suggests a much more targeted set of interventions than 16,000.

Proliferation of discoveries

Despite all the infrastructural cross-validation and regularization of plural expression, machine learning does not stabilize genomes as data objects. In many ways, it gives rise to further transformations and variations, new sources of error.[11] If, on the one hand, machine learners offer to regularize transient multiplicities (such as gene expression in complex disorders), then, on the other hand, within genomics, machine learning exhibits considerable epistemic instability that needs to be regulated.

For instance, since 2003, the U.S. Food and Drug Administration has conducted a study of data analysis techniques for microarray data:

> The US Food and Drug Administration MicroArray Quality Control (MAQC) project is a community-wide effort to analyze the technical performance and practical use of emerging biomarker technologies (such as DNA microarrays, genome-wide association studies and next generation sequencing) for clinical application and risk/safety assessment (Parry et al. 2010, 292).

Phase I of the U.S. Federal Drug Administration-led MAQC addressed many issues of data analysis in the context of the clinical applications of gene expression analysis using microarrays. The primary statistical issue there was minimizing the "false discovery rate" (Slikker 2010, S1), a typical biostatistical problem linked to false positives. In its

11 One important difficulty is the increasingly visible presence of variations in genomes. These variations first become visible after the assembly of whole genome sequences. Genomes of individuals of the same species vary in having slightly different versions of the genes (alleles), many of which differ only by single nucleotide base pairs. Whole genome sequencing made these differences, known as single nucleotide polymorphisms (SNPs), much more apparent. They occur in their tens of millions in the human genome (some 100 million are reported in the NCBI dbSNP database). In coding regions, SNPs may occasion changes in protein structure; in non-coding regions, that can affect how gene expression occurs or is regulated. Like genes, SNPs can be detected using microarrays. SNP microarrays are commonly used in genome-wide association studies (GWAS) that explore complex genetic traits and conditions. SNP-based DNA microarrays measure the occurrence of millions of SNPs in a given biological sample.

Choose k, a positive integer which is large but small compared to the sample sizes. Specify a metric in the sample space, for example ordinary Euclidean distance. Pool the two samples and find, of the k values in the pooled samples which are nearest to z, the number M which are X's. Let $N = k-M$ be the number which are Y's. Proceed with the likelihood ratio discrimination, using however $\frac{M}{m}$ in place of $f(z)$ and $\frac{N}{n}$ in place of $g(z)$. That is, assign Z to F if and only if

$$\frac{M}{m} < c\,\frac{N}{n}.$$

Figure 7.4
The earliest formulation of the *k*-nearest neighbors model from Evelyn Fix and Joseph Hodges' work (1951).

second phase known as MAQC-II starting in 2007, however, the focus rested on the construction of predictive models for "toxicological and clinical endpoints . . . and the impact of different methods for analyzing GWAS data" (Slikker 2010, 2). On both the clinical and GWAS fronts, the 36 participating research teams tried out many predictive classifier models.

In the shift from MAQC-I to MAQC-II, the problem of variations in the predictions produced by the machine learning models moved to center stage. The problem of variation arises not because any of the different modeling strategies used in machine learning gene expression datasets are wrong or erroneous, but because every model transforms the "feature space" (Parry et al. 2010, 292) in a different way (as we saw in chapter 6 in discussions of different treatments of dimensionality). In the MAQC-II consortium, teams were tasked to build "classifiers" to predict whether a given sample or case belongs to a "normal" or "disease" group. The most popular classifier in the MAQC consortium was the *k* nearest neighbors model: "[a]mong the 19,779 classification models submitted by 36 teams, 9742 were k-nearest neighbor-based (KNN-based) models (that is, 49.3% of the total)" (Parry et al. 2010, 293). But these models varied greatly in their predictions: "there have been large variations in prediction performance among KNN models submitted by different teams" (293). The genome itself varies, and the population of machine learners shows variations.

What accounts for this variation? First of all, the teams did not build single models. As is the norm in machine learning, they iterated over thousands. In their attempt to normalize the variations of their models, one of the research groups in MAQC-II write, "for clinical end points and controls from breast cancer, neuroblastoma and multiple myeloma, we systematically generated 463,320 *k*-nn [*k*-nearest neighbor] models by varying feature ranking method, number of features, distance metric, number of neighbors, vote weighting and decision threshold" (Parry et al. 2010, 292). A striking proliferation of models on a population scale strives to tame the variations of predictive models. The number of predictive models constructed here rivals the number of SNPs typically assayed by the microarrays. It seems as if the dimensions of the data as well as the population of models have vastly increased. This population exhibits many of the problems of variation, irregularity, transience, and plurality found in the genomic referential.

Variations in the object or in the machine learner?

"The method of k-nearest neighbors makes very mild structural assumptions: its predictions are often accurate but can be unstable" write Hastie and co-authors (Hastie, Tibshirani, and Friedman 2009, 23). The algorithm, first described by Evelyn Fix and Joseph Hodges working at Berkeley in the early 1950s (Fix and Hodges 1951), is extremely simple in mathematical terms.[12] Equation 7.3 shows almost the entire algorithm:

$$\hat{Y}(x) = \frac{1}{k} \sum_{x_i \in N_k(x)} y_i \qquad (7.3)$$

"where $N_k(x)$ is the neighborhood of x defined by the k closest points x_i in the training sample" (Hastie, Tibshirani, and Friedman 2009, 14). The algorithm effectively takes the average values of points in the neighborhood N_k and uses that value to predict the result (a classification or prediction) for a given point or instance. As Hastie and co-authors put it, the neighborhood is just those k points near the case

12 Fix and Hodges frame their suggestion of the *k*-nearest neighbor model in this way: "there seems to be a need for discriminative procedures whose validity does not require the amount of knowledge implied by by the normality assumption, the homoscedastic assumption, or any assumption of parametric form. The present paper is, as far as the authors are aware, the first one to attack subproblem (iii): can reasonable discrimination procedures be found which will work even if no parametric form can be assumed?" (Fix and Hodges 1951, 7). Subproblem (iii) in this quote refers to the challenge of deciding which of two populations an observed case belongs to if we know nothing about the parameters describing the two populations.

under consideration. The assumption here, as in nearly all machine learners transforming the vector space, is that proximity in vector space implies similarity in class or grouping. This assumption was formally described in the late 1960s in another highly cited paper (Cover and Hart 1967) on "Nearest Neighbor Pattern Classification." Neighboring points in the vector space are more similar than those at a distance. As equation 7.3 shows, *k*-nearest neighbors seems to have only one parameter, the value *k*, the number of neighbors that a given model includes in its definition of a "neighborhood." In contrast to the linear forms of the models (formulated in equations 7.2 or 7.1), equation 7.3 seems to require little training, supervision, or regularization to work as a classifier. Although nearly all of the models discussed in this and earlier chapters work with a smooth functional form of the line or curve as their basic way of transforming vector space, *k*-nearest neighbors generate highly nonlinear boundaries wending their way through the data. Because they are not guided by parameters (apart from the value of the hyperparameter *k*), these boundaries can be unstable.

Even when data belong to two classes (e.g., `normal` vs. `not-normal`), decision boundaries produced by *k*-nn can be unstable. The example in figure 7.5 shows two models, one for $k = 5$ and the other for $k = 2$. Each model examines the relations between 2, 5 points in deciding whether a particular case belongs to one class or another. Although *k*-nn constructs local clusters and traces out an irregular decision boundary, this classificatory power comes at the cost of instability. (This is another version of the bias-variance decomposition of machine learner errors discussed in chapter 5.)

More or wider data exacerbate the instability. As dimensions or features in the dataset increase, the local neighborhood needed to capture a fraction of the volume of the data expand. It becomes more likely that most sample points will lie close to the boundary of the sample space, where they will be affected by the neighboring space. The result is that "in high dimensions all feasible training samples sparsely populate the input space" (Hastie, Tibshirani, and Friedman 2009, 23). Because *k*-nn allows for nonlinear interactions between features, for instance, small differences in the number of points in particular neighborhoods can drastically affect some stretches of the boundaries (as we see in comparing the right- and left-hand plots in figure 7.5). These kinds of topological instability account for the propensity of machine learning treatments of feature-rich genomic data to produce accurate but unstable predictions. We can begin to see how a MAQC-II team might have produced 463,000 *k*-nearest neighbor models in an effort to normalize and regulate predictive predictions. The price of accurate predictivity in genomics is variation in prediction.

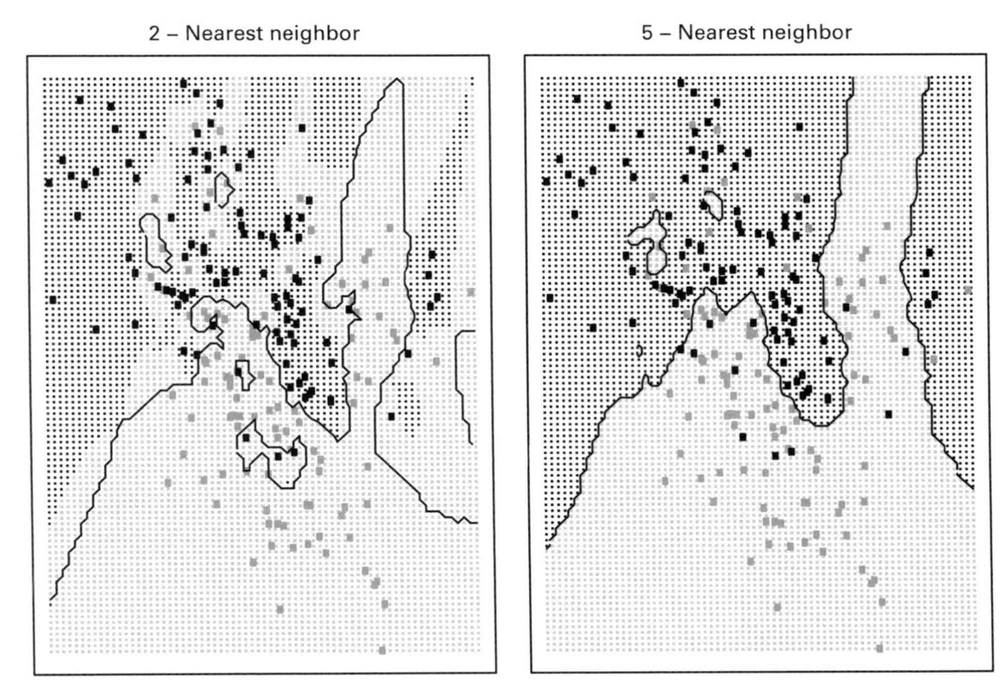

Figure 7.5
k-nn models for simulated data in two classes. The decision boundary that separates the two classes is non-linear for both versions of the k-nn model. For $k = 5$, the decision boundary is more non-linear than for $k = b$.

Whole genome functions

Cores, microarrays, SNPS, and many models, infrastructural scaling, biological variation, and the populations of unruly machine learners entwine in referential entanglement. If genomes are scientific hyperobjects (with epistemic, speculative, financial, and biopolitical resonances), then what part does their referential cross-validation with machine learning play in the transformation of knowledge?

Genomic data—beginning with DNA sequences, then levels of gene expression, followed by genome-wide association studies of small mutations—have been a constant $p \gg N$ antigen in machine learning. Techniques of regularization—the lasso—of linear models discussed in this chapter came to light and were first demonstrated on genomic data produced in the mid-1990s. Throughout the ongoing development and enrichment of DNA and protein sequencing techniques, replete with a vast and quite dynamic bioinformatic infrastructure, machine learning and genomics have been cross-validating in practice. Scientists, statisticians, datasets, and machine learners traffic between genomics and machine learning at almost every level, ranging from the sequence assembly to testing and analysis of DNA data in clinical settings. In genomics, elementary practices of aligning and assembling sequences into whole genomes were reconfigured probabilistically through machine learning models.

Almost every subsequent development in genomics (and related fields such as proteomics) follows a referential flow of materializing transcription, infrastructural cross-validation, and regularizing differentiation. An entity whose constitution is thoroughly dependent on prediction or algorithmic classification displays variations and grouping (such as gene expression, the linkage disequilibrium of SNPs, the seeming abundance of junk DNA that is actually functional, etc.) that attract further efforts to differentiate, regularize, and classify ever more subtly distributed differences. Elementary practices in contemporary genomics such as sequence alignment were explicitly formulated as generative models to be constructed using algorithms such as expectation maximization. As we see in the vignettes from *Elements of Statistical Learning*, the demonstration of Google Compute cloud computing, or for that matter in the myriad publications in both machine learning and life science journals that make use of support vector machines, neural networks, linear discriminant analysis, or random forests, machine learners reconfigure the conduct of scientific research and construct new kinds of statements, new types of objects (genomes in particular are difficult to conceive without their probabilistic modeling), and, as will be discussed in the next chapter, subject positions (bioinformaticians, computational biologists, data scientists, and others).

What is at stake for machine learners in a scientific setting like genomics? Foucault writes, "We should distinguish carefully between *scientific domains* and *archaeological territories*" (Foucault 1972, 183). Knowledge stems from the practices that connect objects, field, subjects, statements, and institutions. Sciences are always localized within a field of knowledge that may exceed and mutate in ways that alter local sciences. Science, for Foucault and perhaps for science studies more generally, needs knowledge practices that exceed, surround, and indeed do something different to science. Machine learning is just such an operational formation.

Could we pose or address any normative questions by becoming aware of and articulating machine learning with science with greater clarity? Genomic science, in its cross-validation with machine learning, displays some of the tendencies to reduce divergences and corral differences typical of knowledge economies more generally. The philosopher of science, Isabelle Stengers, writes:

> with the knowledge economy, we may have scientists at work everywhere, producing facts with the speed that new sophisticated instruments make possible, but that the way those facts are

interpreted will now mostly follow the landscape of settled interests.... We will more and more deal with instrumental knowledge (Stengers 2011, 377).

As we see in the 600,000 cores of Google Compute applied to exploration of associations in cancer genomics using random forests or the lasso applied to microarray SNP data, machine learning rapidly produces facts. Stengers suggests that the risk here is that divergence and unexpected forms of experimental result are somewhat diminished as a result. Machine learning in genomics might produce a self-organizing map that poses questions following the "landscape of settled interests" or the status quo.

I see matters as slightly more complicated than an instrumental production of knowledge. In the biosciences of the last two decades, machine learning seeks to disaggregate, compartmentalize, and rank those aspects of genomes—their confused variations, their manifold spatial and temporal relationality in biological processes—that seem most distant and difficult to derive from putatively linear, monolithic, and searchable DNA sequence data. DNA can be laid down in tracks or grids, aligned and annotated in uniquely addressable database records, but the problem of how this extensive vector maps onto the subtle, pervasive, and transient forms of temporal and spatial reshaping in life forms remains. None of the examples of genomic data in *Elements of Statistical Learning* uses whole genomes. In the feature-rich spaces countenanced by machine learning, we see attempts to embed manifolds in local regions, local linearities. Sometimes these local regions are regions of annotated DNA or nonlinear interactions between sets of genes, as in the GWAS analysis of epistasis. At other times, these local regions are forms of life in a more general sense—clinical outcomes or diagnostic tests—as in MAQC-II.

The enunciative function of machine learning inscribes the possibility of genomes as multitemporal, interconnected expressions of variation. Such regularizing and potentializing of things on new infrastructural, collective, and domain-specific scales outstrips instrumental purposes. In *The Archaeology of Knowledge*, Foucault describes discourse as "controlled, selected, organised and redistributed according to a certain number of procedures, whose role is to avert its powers and its dangers, to cope with chance events, to evade its ponderous, awesome materiality" (Foucault 1972, 216). Something similar flows through operational formations such as machine learning in their entanglements with sciences. Controlling, selecting, and organizing, it almost inadvertently affirms a ponderous, awesome materiality of data practice.

8 Propagating Subject Positions

If a proposition, a sentence, a group of signs can be called "statement," it is not therefore because, one day, someone happened to speak them or put them into some concrete form of writing; it is because the position of the subject can be assigned (Foucault 1972, 95).

Generalization error is what we care about (Ng 2008h).

Predict if an online bid is made by a machine or a human, "Facebook Recruiting IV: Human or Robot?" (Kaggle 2015d).

Who is the machine learner subject? In early 2002, while carrying out an ethnographic study of "extreme programming," a software development methodology popular at that time (Mackenzie and Monk 2004), I spent several months visiting a company in Manchester developing software for call centers. The software was to manage "knowledge" in call centers such that any query from a caller could be readily answered by staff who would query a knowledge management system to find answers to the query. This system was marketed on the promise of machine learning. It relied on an artificial neural network that learned to match queries and responses over time. A taciturn neural network expert, Vlad, sat in a different part of the room from the developers working on the databases and web interfaces. Vlad's work on the neural network was at the core of the knowledge management system yet outside the orbit of the software development team and its agile software development processes. The rest of the team generally regarded Vlad and the neural net as an esoteric, temperamental, yet powerful component, a hidden node, we might say, of the knowledge management system.

As we have seen with `kittydar`, the position of machine learners is changing. They are no longer exotic or specialized, but banal or occasionally spectacular. Hilary Mason, who was Chief Scientist at bitly.com (an online service that shortens URLs), outlined an everyday machine learning subject position at a London conference in 2012 called "Bacon: Things Developers Love":

You have all of these things that are different—engineering, infrastructure, mathematics and computer science, curiosity and an understanding of human behavior—that is something that usually falls under the social sciences. At the intersection of all these things are wonderful people. But we're starting to develop something new, and that is - not that all of these things have not been done for a very long time - but we are only just now building systems where people, individual people, have all of these tools in one package, in one mind (Hilary Mason, Chief Scientist, bitly.com) (Mason 2012).

These "things that are different," what I have been calling an operational formation, assign subject positions. In what ways does machine learning assign subject positions? In front of an audience of several hundred software developers, Mason describes shifts in the work of programming associated with the growth of large amounts of data associated with human behavior. At the center of this shift stand "wonderful people" who combine practices and knowledge of communication infrastructure, technology, statistics, and human behavior through curiosity and technical skills. Mason was, in effect, telling her audience of software developers who they could become in relation to expansive changes occurring in and around their work. The title of her talk was "Machine Learning for Hackers," and her audience were hackers or programmers whose coding and programming attention may have been previously trained on web interfaces or database queries but was now drawn toward machine learning. A change in programming practice and a shift toward machine learning was, she implied, the key to programmers becoming the wonderful people, agents of their own time, capable of doing what is only now just possible because it is all together in "one package, one mind."

Neural nets stand at an intersection of infrastructure and cognition and then propagate subject positions forward and backward. Their operations encourage and elicit competitively ranking as an ordering that compares human and machines, as well as subject positions more generally.

Propagation across human–machine boundaries

The concatenation of "one package, one mind" does not definitively allocate agency to people or things. (A "package," after all, is another name for a library of code.) Mason adumbrates the outline of a subject position located at the intersection of network infrastructure, mathematics, and human behavior.[1] Mason, herself one of *Fortune* magazines "Top 40 under 40" business leaders to watch (CNN 2011) and also featured in

1 In earlier work on machine learning (Mackenzie 2013b), I presented programmers as agents of anticipation, suggesting that the turn to machine learning among programmers could be useful in understanding how predictivity was being done amid broader shift to the regime of anticipation described by Vincenne Adams, Michelle Murphy, and Adele Clarke (Adams, Murphy, and Clarke 2009). Subsequent developments in machine learning, even just in the last

Glamour, a teenage fashion magazine (Mason 2012), might personify such a wonderful person. She is not a lone example. In mid-2016, Google announced a comprehensive program to retrain its software developers as machine learners (Levy 2016).[2]

"It is the privileged machine in this context that creates its marginalized human others," writes Lucy Suchman in her account of the encounters that "effect 'persons' and 'machines' as distinct entities" (Suchman 2006, 269). Despite the fact that Mason and other relatively well-known human machine learners are not exactly marginalized (just the opposite; they achieve minor celebrity status in some cases), Suchman recommends "recognition of the particularities of bodies and artifacts, of the cultural-historical practices through which human–machine differences are (re-)iteratively drawn, and of the possibilities for and politics of redistribution across the human machine boundary" (285). The intersections that machine learners currently occupy are heavily redistributional. In almost every instance, machine learners claim to do something that humans alone, no matter how expert, could not. Does the redistribution of engineering, mathematics, curiosity, infrastructure, and "something that usually falls under the social sciences" (but perhaps no longer does so?) both energize subjects ("its a pretty exciting time to be in any of these things") and assign them a marginal albeit still pivotal position in relation to privileged machines?

Machine learner subject positions are the topic of this chapter. I concentrate on artificial neural networks, or neural nets, in their various forms, ranging from the multilayer perceptron (MLP) to the convolutional neural nets (CNN), recurrent neural nets (RNN) and deep belief networks of many recent deep learning projects (particularly in machine learning competitions, as discussed below (Dahl 2013)), in exploring the redrawing of subject-machine positions. Neural nets propagate among infrastructures, engineering, and human behavior (as Mason puts it), redrawing human–machine differences, sometimes making it hard to see what subject position they entail, where subjects are located, or what they say, see, and do.

Like other machine learners, neural nets redraw human–machine differences. Geoffrey Hinton, Simon Osindero, and Yee-Why Teh, writing in *Neural Computation* in 2006,

---

three years, confirm that view, but in this chapter and in this book more generally, I focus less on transformations in programming practice and software development and more on the asymmetries of different machine learner subjects in relation to infrastructures and knowledge.

2 Other figures we might follow include Claudia Perlich, Andrew Ng, Geoffrey Hinton, Corinna Cortez, Daphne Koller, Christopher Bishop, Yann LeCun, or Jeff Hammerbacher. Although some women's names appear here, in any such list, men's names are much more likely to appear. This is no accident.

described a "fast learning algorithm for deep belief nets" (Hinton, Osindero, and Teh 2006). Their description, although mostly couched in terms of conditional distributions, model parameters, and error rates, also contains a section titled "Looking into the Mind of a Neural Network" (1545–1546). In that section, they describe how they used their deep belief network to *generate* rather than classify images.[3] In the process, they were able to see what the "associative memory has in mind" (1545). The term "mind," they comment, "is not intended to be metaphorical" (1546) because the neural net in question has a distributed memory of the digits it has seen. Put slightly more formally, "the network has a full generative model, so it is easy to look into its mind—we simply generate an image from its high-level representations" (1529). "Looking," as is often the case in machine learning, takes the form of diagramming a pattern, partition, or strain in the data.

The substitution of mind for model resonates with certain aspects of the field of artificial intelligence (which has typically relied on rule-based or symbolic reasoning), but the appearance of mind in the form of a generative model (see chapter 5) suggests a rather different subject position. Archaeologically, the description of machine learner subjects first of all entails locating the multiple positions linked to different groupings and statements in the operational formation. As I will suggest, neural nets are particularly interesting because they redraw human–machine boundaries through a combination of feeding-forward of potentials and propagating backward of differences specifically concerned with images. Similarly, the practice of machine learning shifts subject positions in a backward and forward movement. It propagates potentializing optimism even as it undercuts the very differences that underpin that optimism.

Almost every machine learning class, textbook, demonstration, and, in recent years, machine learning competition at some point turns to neural nets. Neural nets display, however, some instability in the research literature. Figure 8.1 shows the most frequent keywords for technical publications across the three main disciplinary domains inhabited by machine learners. Although neural nets rank high in computer science and engineering disciplines (appearing just after support vector machines), they do not appear in the statistics literature until the mid-2000s. The prominence of neural nets on the engineering side of machine learning suggests a specific enunciative mapping.

Neural nets are often described from a deeply split perspective. At some moments, the description turns toward human subjects or, at least, the brains of human subjects.

3 In the case of this paper and many others related to neural nets, the images are of hand-written digits. These digits have an almost constitutive role, as I discuss in this chapter.

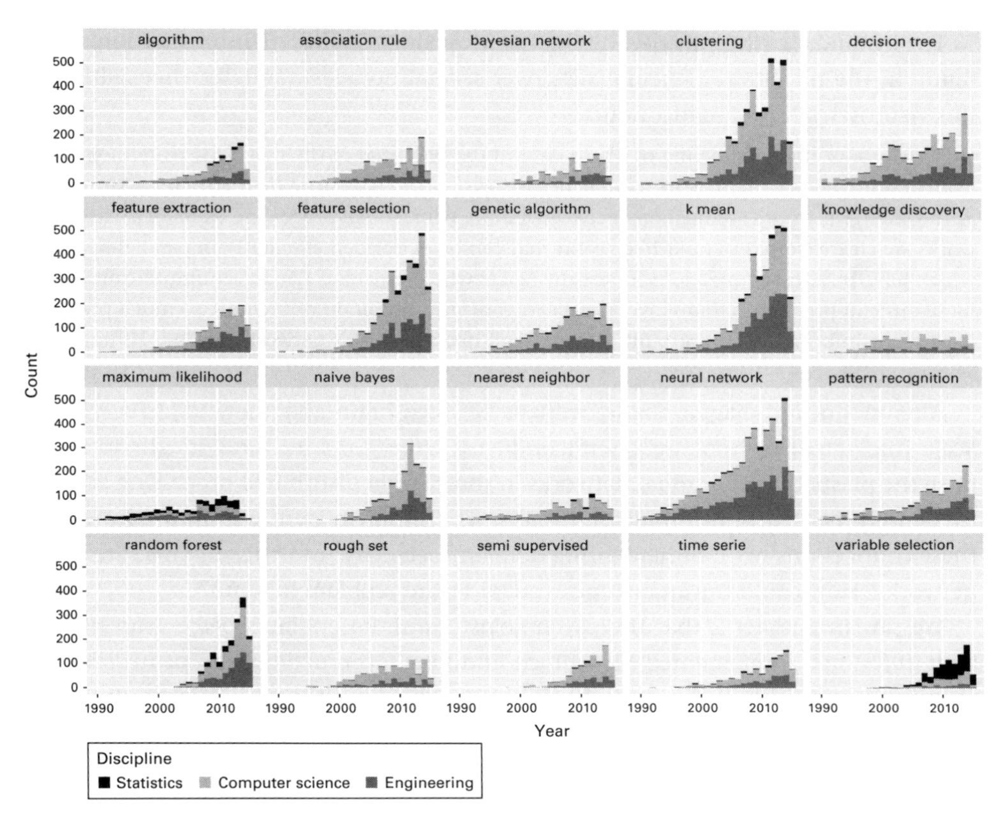

Figure 8.1
Techniques and concepts most frequently mentioned in machine learning publication keywords, 1955–2015.

At other moments, neural nets turn toward the vectorization of data. Neural nets constantly oscillate between brain and information infrastructure. In some ways, they renew longstanding cybernetic hopes of bring brains and computation together in models of computational intelligence and agency (Hayles 1999). Although they stem from a biological inspiration (dating at least back to the work by McCulloch and Pitts in the 1940s; Halpern 2015; Wilson 2010), they gained traction first in the 1980s and then again from the mid-2000s onward, as ways of dealing with changing computational infrastructures and the difficulties of capitalizing on infrastructure that is powerful but hard to control. In the course of 50 years, their serial reinvention—from perceptron via neural net to deep belief net—triply redistributes subject positions amid infrastructural reconfigurations and vectorization.

For instance, writing in the 1980s, David Ackley, Geoffrey Hinton (an important figure in the inception of neural nets over several decades), and Terrence Sejnowski link neuroscience and semiconductors:

> Evidence about the architecture of the brain and the potential of the new VLSI technology have led to a resurgence of interest in "connectionist" systems . . . that store their long-term knowledge as the strengths of the connections between simple neuron-like processing elements. These networks are clearly suited to tasks like vision that can be performed efficiently in parallel networks which have physical connections in just the places where processes need to communicate. . . . The more difficult problem is to discover parallel organizations that do not require so much problem-dependent information to be built into the architecture of the network (Ackley, Hinton, and Sejnowski 1985, 147–148).

Alignments and diagrammatic overlaps between brain and "new VSLI [Very Large Scale Integrated] technology"—semiconductor chips—architectures sought to reproduce the plasticity of neuronal networks in the parallel distributed processing enabled by densely packed semiconductor circuits. The problem here was how to organize these connections without hardwiring domain specificity into "the architecture of the network." How could the architectures adapt to the problem in hand?

We saw in chapter 2 that the psychologist Frank Rosenblatt's perceptron (Rosenblatt 1958) first implemented McCulloch and Pitts' cybernetic vision of neurones as models of computation (Edwards 1996). Although computer science research on the perceptron wilted under criticism from artificial intelligence experts such as Marvin Minsky (Minsky famously showed that a perceptron cannot learn the logical exclusive OR or `XOR` function; Minsky and Papert 1969), cognitive psychologists such as David Rumelhart, Geoffrey Hinton, and Ronald Williams persisted with perceptrons, seeking to generalize their operations by connecting them together in networks (also known as multilayer perceptrons). In the mid-1980s, they developed the back-propagation algorithm (Rumelhart, Hinton, and Williams 1985; Hinton 1989), a way of optimizing the connections—known as weights or parameters—between nodes (neurones) in the network in response to features in the data (see figure 8.2).

The back-propagation algorithm directly addressed the problem of learning to modify network organization without reliance on problem-specific architectures or hand programming. Effectively, it constructs an architecture of generalization. Although cognition and the idea that machines would be cognitive (rather than, say, mechanical, calculative, or rule-based) mesmerized research work in artificial intelligence for several decades, the development of the back-propagation algorithm as a way for a set of connected computational nodes to learn came with explicit infrastructural resonances.

Unclassified
SECURITY CLASSIFICATION OF THIS PAGE

REPORT DOCUMENTATION PAGE

1a. REPORT SECURITY CLASSIFICATION
Unclassified

1b. RESTRICTIVE MARKINGS

2a. SECURITY CLASSIFICATION AUTHORITY

3. DISTRIBUTION/AVAILABILITY OF REPORT
Approved for public release;
distribution unlimited.

2b. DECLASSIFICATION/DOWNGRADING SCHEDULE

4. PERFORMING ORGANIZATION REPORT NUMBER(S)
ICS 8506

5. MONITORING ORGANIZATION REPORT NUMBER(S)

6a. NAME OF PERFORMING ORGANIZATION
Institute for Cognitive
Science

6b. OFFICE SYMBOL
(If applicable)

7a NAME OF MONITORING ORGANIZATION

6c. ADDRESS *(City, State, and ZIP Code)*
C-015
University of California, San Diego
La Jolla, CA 92093

7b. ADDRESS *(City, State, and ZIP Code)*

8a. NAME OF FUNDING/SPONSORING ORGANIZATION
Personnel & Training Research

8b. OFFICE SYMBOL
(If applicable)

9. PROCUREMENT INSTRUMENT IDENTIFICATION NUMBER
N00014-85-K-0450

8c. ADDRESS *(City, State, and ZIP Code)*
Code 1142 PT
Office of Naval Research
800 N. Quincy St., Arlington, VA 22217-5000

10. SOURCE OF FUNDING NUMBERS

PROGRAM ELEMENT NO.	PROJECT NO.	TASK NO.	WORK UNIT ACCESSION NO
	NR 667-548		

11. TITLE *(Include Security Classification)*
Learning Internal Representations by Error Propagation

12. PERSONAL AUTHOR(S)
David E. Rumelhart, Geoffrey E. Hinton, and Ronald J. Williams

13a. TYPE OF REPORT
Technical

13b. TIME COVERED
FROM Mar 85 TO Sept 85

14. DATE OF REPORT *(Year, Month, Day)*
September 1985

15. PAGE COUNT
34

16. SUPPLEMENTARY NOTATION **To be published in J. L. McClelland, D. E. Rumelhart, & the PDP Research Group, *Parallel Distributed Processing: Explorations in the Microstructure of Cognition: Vol 1. Foundations.* Cambridge, MA: Bradford Books/MIT Press.**

17. COSATI CODES

FIELD	GROUP	SUB-GROUP

18. SUBJECT TERMS *(Continue on reverse if necessary and identify by block number)*
learning; networks; perceptrons; adaptive systems;
learning machines; back propagation

19. ABSTRACT *(Continue on reverse if necessary and identify by block number)*

This paper presents a generalization of the perceptron learning procedure for learning the correct sets of connections for arbitrary networks. The rule, called the generalized delta rule, is a simple scheme for implementing a gradient descent method for finding weights that minimize the sum squared error of the system's performance. The major theoretical contribution of the work is the procedure called error propagation, whereby the gradient can be determined by individual units of the network based only on locally available information. The major empirical contribution of the work is to show that the problem of local minima is not serious in this application of gradient descent.

20. DISTRIBUTION/AVAILABILITY OF ABSTRACT
☒ UNCLASSIFIED/UNLIMITED ☐ SAME AS RPT. ☐ DTIC USERS

21. ABSTRACT SECURITY CLASSIFICATION
Unclassified

22a. NAME OF RESPONSIBLE INDIVIDUAL

22b. TELEPHONE *(Include Area Code)*

22c. OFFICE SYMBOL

DD FORM 1473, 84 MAR

83 APR edition may be used until exhausted.
All other editions are obsolete.

SECURITY CLASSIFICATION OF THIS PAGE
Unclassified

Figure 8.2
An early publication of the back-propagation algorithm (Rumelhart 1985).

The resonances between computational architectures and human cognition (centered on vision) became much more palpable around 2006, when "deep belief nets" appeared as a way of training many-layered neural nets implemented on much larger computational platforms (Hinton, Osindero, and Teh 2006). These resonances continue to echo today and indeed attract much attention.[4] Like the advent of VSLI in the early 1980s, the vast concentrations of processing units in contemporary data centers (hundreds of thousands of cores as we saw in the case of Google Compute in the previous chapter 7) and in the graphics cards developed for high-end gaming and video rendering (GPUs for PC gaming now typically have a thousand and sometimes several thousand cores) pose the problem of organizing infrastructure so that processes can communicate with each other. Machine learners have become important as mutable infrastructural elements as well as epistemic instruments.

Oscillating between cognition and infrastructures, between people and machines, neural nets suggest a way of thinking not only about how "long-term knowledge" takes shape today but about subject positions associated with machine learning. As infrastructural reorganization takes place around learning, and around the production of statements by machine learners, both human and nonhuman machine learners are assigned new positions. These positions are sometimes hierarchical and sometimes dispersed. The machine learner subject position is mobile rather than a single localized form of expertise (as we might find in a clinical oncologist, biostatistician, or geologist). Because machine learners vectorize, optimize, probabilize, differentiate, and refer, what counts as agency, skill, action, experience, and learning shifts constantly. It is intimately bound and connected to transforms in infrastructure, variations in referentiality (such as we have seen in the construction of the vector space), and competing forms of

4 Although mainstream media accounts of machine learning are not the focus of my interest here, it is hard to ignore the extraordinary level of interest that deep learning projects and techniques have attracted in the last few years. Articles have appeared in all the usual places: *The New York Times* (Markoff 2012), *Wired* (Garling 2015), or *The Guardian* (Arthur 2015). In many of these accounts, machine learning and neural nets in particular appear both in the guise of the existential threat of artificial intelligence and as a mundane device (e.g., speech recognition on a mobile phone). The spectacular character of deep learning could be analyzed in terms like that of the genomes discussed in chapter 7. In both cases, the advent and transformation of these machine learners is closely linked to networked platforms (such as Google, Facebook, Yahoo, and Microsoft) and their efforts to encompass within their services as many elements of experience, exchange, communication, and power as possible. Deep learning machine learners currently focus mostly on images (photographs and video) and sounds (speech and music), and they usually attempt to locate and label objects, words, or genres.

accumulation or positivity. As Suchman suggests, examining privileged machines such as neural nets is a way to pay attention to the dispersed and somewhat disconnected sites from which subjects program, observe, design, and respond to machine learners.

Competitive positioning

How do neural nets come to oscillate between different subject positions? The ranking of keywords in figure 8.1 suggests that machine learning as an operational knowledge formation attributes a privileged and constitutive function to neural nets. Neural nets concurrently spread into many difference disciplines: cognitive science, computer science, linguistics, adaptive control engineering, psychology, finance, operations research, and so on, and particularly statistics and computer science during the 1980 and 1990s. This dendritic growth did not just popularize machine learning. It brought engineering and statistics together more strongly. Ethem Alpaydin, a computer scientist, writes:

> Perhaps the most important contribution of research on neural networks is this synergy that bridged various disciplines, especially statistics and engineering. It is thanks to this that the field of machine learning is now well established (Alpaydin 2010, 274).

The forms of this field-making bridging are various.[5] The primary meeting point of different disciplines has perhaps been the machine learning competitions of the 1990s, which pitted neural nets against other machine learners, such as the support vector machine. Many of these competitions focused on vision-related problems such as recognizing handwritten numerals. The handwritten digits used in these competitions, particularly the Neural Information Processing System workshops and KDD Cup (Knowledge Discovery and Data Mining) (KDD 2013), all come from the `mnist` dataset, and during the 1990s, much effort focused on crafting neural nets to recognize these 60,000 or so handwritten digits.

Elements of Statistical Learning devotes a lengthy section to the analysis of image recognition competitions that began in the early 1990s and continue today. Like Alpaydin, it affirms the coordinating effect of these competitions on the development of machine learning:

> This problem captured the attention of the machine learning and neural network community for many years, and has remained a benchmark problem in the field (Hastie, Tibshirani, and Friedman 2009, 404).

5 We saw some use of neural nets in genomics in the previous chapter (7). The initial publication of the `SRBCT` microarray dataset (Khan et al. 2001) relied on neural nets.

As Hastie and co-authors observe, "at this point the digit recognition datasets became test beds for every new learning procedure, and researchers worked hard to drive down the error rates" (Hastie, Tibshirani, and Friedman 2009, 408–409). During the 1990s, zipcodes on envelopes (the set of handwritten digits we have already seen in chapter 6, the `mnist` datasets; LeCun and Cortes 2012) became a primary focus of learning. The many competitions focused on the `mnist` dataset are, I suggest, a form of demonstration and testing of machines and people that propagate machine-human differences in machine learning.

Although they brought statistics and computer science closer, neural networks have had a somewhat problematic position in machine learning. Even in relation to the paradigmatic handwritten digit recognition problem, neural nets struggled to gain purchase precisely because a human subject position remained intimately interwoven into their operation. On the one hand, their analogies and figurations as sophisticated neuronal-style models suggested cognitive capacities surpassing the more geometrical, algebraic, and statistically grounded machine learners such as linear discriminant analysis, logistic regression, or decision trees. On the other hand, the density and complexity of their connective architecture made them difficult to train. Neural nets could easily overfit the data. As *Elements of Statistical Learning* puts it, it required "pioneering efforts to handcraft the neural network to overcome some these deficiencies. . . , which ultimately led to the state of the art in neural network performance" (Hastie, Tibshirani, and Friedman 2009, 404). It is rare to find the word "handcraft" in machine learning literature. The operational premise of most machine learners is that machine learning works without handcrafting or that it automates what had previously been programmed by hand. Somewhat ironically, the competition to automate recognition of handwritten digits, the traces that epitomize movements of hands, entailed much handcrafting and recognition of variations in performances of the machine.

The unstable position of subjects in relation to neural nets are frequently discussed in contrasting terms by machine learners.[6] They often point to a transformation in the work of machine learning:

> Neural networks went out of fashion for a while in the 90s - 2005 because they are hard to train and other techniques like SVMs beat them on some problems. Now people have figured out better methods for training deep neural networks, requiring far fewer problem-specific tweaks. You

6 Neural nets also receive uneven attention in the machine learning literature. In Andrew Ng's Stanford CS229 lectures from 2007, they receive somewhat short shrift: around 30 minutes of discussion in Lecture 6, in between Naive Bayes classifiers and several weeks of lectures on support vector machines (Ng 2008c). As he introduces a video of an autonomous vehicle steered by a neural net after a 20-minute training session with a human driver, Ng comments that, "neural nets

can use the same pretraining whether you want a neural network to identify whose handwriting it is or if you want to de-cipher the handwriting, and the same pretraining methods work on very different problems. Neural networks are back in fashion and have been outperforming other methods, and not just in contests (Zare 2012).

The somewhat vacillating presence of neural nets in the machine learning literature finds parallels in the movements of individual machine learners. Yann Le Cun's work on optical character recognition from the 1980s to the 1990s is said to have discovered the back-propagation algorithm at the same time as Rumelhart, Hinton, and Williams (Rumelhart, Hinton, and Williams 1986). His implementations in `LeNet` won many research machine learning competitions during the 1990s. In 2007, Andrew Ng could casually observe that neural nets *were* the best, but in 2014, Le Cun found himself working on machine learning at Facebook (Gomes 2014). Similarly, the cognitive psychologist Geoffrey Hinton's involvement in the early 1980s work on connectionist learning procedures in neural nets and subsequently on "deep learning nets" (Hinton and Salakhutdinov 2006) delivered him to Google in 2013.

---

were the best for many years." The lectures quickly move on to the successor, support vector machines. In *Elements of Statistical Learning*, a whole chapter appears on the topic but is prefaced by a discussion of the antecedent statistical method of "projection pursuit regression." The inception of "projection pursuit" is dated to 1974, and thus precedes the 1980s work on neural nets that was to receive so much attention. In *An Introduction to Statistical Learning with Applications in R*, a book whose authors include Hastie and Tibshirani, neural nets are not discussed and indeed not mentioned (James et al. 2013). Textbooks written by computer scientists such as Ethem Alpaydin's *Introduction to Machine Learning* do usually include at least a chapter on them, sometimes under different titles such as "multi-layer perceptrons" (Alpaydin 2010). Willi Richert and Luis Pedro Coelho's *Building Machine Learning Systems with Python* likewise does not mention them (Richert and Coelho 2013). Cathy O'Neil and Rachel Schutt's *Doing Data Science* mentions them but does not discuss them (Schutt and O'Neil 2013), whereas both Brett Lantz's *Machine Learning with R* (Lantz 2013) and Matthew Kirk's *Thoughtful Machine Learning* (Kirk 2014) devote chapters to them. In the broader cannon of machine learning texts, the computer scientist Christopher Bishop's heavily cited books on pattern recognition dwell extensively on neural nets (Bishop et al. 1995; Bishop 2006). Among statisticians, Brian Ripley's *Pattern Recognition and Neural Networks* (Ripley 1996), also highly cited, placed a great deal of emphasis on them. But these textbooks stand out against a pointillistic background of hundreds of thousands of scientific publications mentioning or making use of neural nets since the late 1980s in the usual litany of fields—atmospheric sciences, biosensors, botany, power systems, water resource management, internal medicine, and so on. The swollen tide of publication attests to some kind of formation or configuration of knowledge invested in these particular techniques, perhaps more so than others I have discussed so far (logistic regression, support vector machine, decision trees, random forests, linear discriminant analysis, etc.).

Trajectories between academic research and industry are not unusual for machine learners. Many of the techniques in machine learning have been incorporated into companies later acquired by other larger companies. Even if there is no spin-off company to be acquired, machine learners have been assigned key positions in many industry settings. Corinna Cortes, co-inventor with Vladimir Vapnik of the support vector machine, heads research at Google New York. In 2011, Ng led a neural net-based project at Google that had, among other things, detected cats in millions of hours of YouTube videos.[7] In 2014, Ng began work as chief scientist for the Chinese search engine, Baidu, leading a team of AI researchers specializing in "deep learning," the contemporary incarnation of neural nets (Hof 2014). In recent years (2012–2015), work on neural nets has again intensified, most prominently in association with social media platforms, but also in the increasingly common speech and face recognition systems found in everyday services and devices. Many of these neural nets are like `kittydar` but implemented on a much larger and more distributed scale (e.g., in classifying videos on YouTube). In contemporary machine learning competitions, as we will see, neural nets again surface as intersectional machines, redistributing differences between humans and machines.

A privileged machine and its diagrammatic forms

What accounts for the somewhat uneven fortunes of the neural net among machine learners? The unevenness of their performance, from limited curiosity in the late 1960s to handcrafted best-in-class performer in the machine learning image classification competitions of the 1990s, from second best competitor in the late 1990s to the spectacular promise of deep belief networks amid the "awesome materiality" of social media image streams in 2012, suggests that some powerful dynamics or becomings are in play around them. These dynamics are not easily understood in terms of celebrity machine learners (human and nonhuman) suddenly rising to prominent or privileged positions in the research departments of social media platforms.[8] Nor does it make sense to

7 Unlike the cats detected by `kittydar`, the software discussed in the introduction to this book, the Google experiment did not use supervised learning. The deep learning approach was unsupervised (Markoff 2012). That is, the neural nets were not trained using labeled images of cats.

8 In any case, social media and search engines cannot be understood apart from the machine learning techniques that have been thoroughly woven through them since their inception. Hence, *Elements of Statistical Learning* devotes several pages to Google's famous *PageRank* algorithm, describing it as an unsupervised learner (Hastie, Tibshirani, and Friedman 2009, 576–578).

attribute the rising fortunes of the neural net to the algorithms, as if some decisive advance occurred in algorithms.

The algorithms such as back-propagation used in neural nets have not, as we will see, been radically transformed in their core operations since the 1980s, and even then the algorithms (principally gradient descent) were not new. There have been important changes in scale (similar to those described in the previous chapter in the case of the `RF-ACE` algorithm and Google Compute), but as is often the case in machine learning, their proliferation occurs through redistributions of knowledge and infrastructure associated with altered subject positions. Although machine learners in their machine form can be assigned a privileged position in the transformations of knowledge and action, human machine learners are not exactly marginalized, at least in celebrated cases such as Ng, Le Cun, Hinton, and others. Rather, we can see varying subject positions emerging in relation to specific devices and data forms (images, sounds, documents) in specific sites (social media platforms and mobiles devices in particular).

A varying subject position surfaces in the operational architecture of neural nets. Despite differences in diagrammatic form, neural net share much with other machine learners. The language of brain, neurones, and cognition associated with neural net covers over the much more familiar vector space and function-finding optimizations they rely on. Diagrammatic groupings and lines of movement operate in neural nets to generate an expansive architecture aligned through a series of well-established transformations. "The central idea," write Hastie and co-authors, "is to extract linear combinations of the inputs as derived features, and then model the target as a nonlinear function of these features. The result is a powerful learning method, with widespread applications in many fields" (Hastie, Tibshirani, and Friedman 2009, 389). The central idea can be seen in the algebraic expressions that Hastie and co-authors provide for the basic neural net model:

$$\begin{aligned} Z_m &= \sigma(\alpha_0 m + \alpha_m^T X) \quad m = 1, \ldots, M \\ T_k &= \beta_0 k + \beta_k^T Z, \quad k = 1, \ldots, K, \\ f_k(X) &= g_k(T), \quad k = 1, \ldots, K, \end{aligned} \tag{8.1}$$

where $Z = (Z_1, Z_2, \ldots, Z_M)$, and $T = (T_1, T_2, \ldots, T_K)$. The activation function $\sigma(v)$ is usually chosen to be the sigmoid $\sigma(v) = 1/(1 + e - v)$ (Hastie, Tibshirani, and Friedman 2009, 392).

Equation 8.1 diagrams some familiar elements as well as some novelty. Some elements of the neural net are already familiar from linear models. The neural networks transform data in a vector space denoted by X. That is common to nearly all machine learners. They make use of the nonlinear sigmoid function that lies at the heart of one of the main linear classifiers used in machine learning, logistic regression. Their training and learning processes have come to rely on the same kinds of cost, loss, or error functions we have seen in other machine learners.

The apparently increasing power of neural nets to learn (to see, find, classify, rank, predict) owes much to diagrammatic substitutions that recombine operations of past machine learners in new intersections. These movements appear in the equations. Equation 8.1 has three lines rather than one, and this layering and its diagonal patterns of indexical referencing running between subscripts distinguish neural nets from the linear models they assemble.

$$\hat{Y} = \hat{\beta}_0 + \sum_{j=1}^{p} X_j \hat{\beta}_j \tag{8.2}$$

Whereas the standard linear model shown in equation 8.2 indexes a single vector space X_j and approximates it using a single function $\hat{Y}$ by searching for the values of the parameters β_j that best incline a flat plane through the data, the three lines shown in equation 8.1 are woven through each other much more consecutively. Each line derives "features" from the line above, adding layers to the network. So-called "hidden layers," such as line two of equation 8.1, repeatedly transform the vector space inside the model. Each node, $1 \ldots K$, adds a new dimension to this internal vector space. In many layered deep learning neural nets, the dimensionality of the vector space vastly expands. Much hinges on the unobstrusive sigmoid function operator written as σ: "a neural network can be thought of as a nonlinear generalization of the linear model, both for regression and classification. By introducing the nonlinear transformation σ, it greatly enlarges the class of linear models" (Hastie, Tibshirani, and Friedman 2009, 394). σ, it seems, allows neural nets to generalize beyond the linear model.[9]

Varying subject positions in code

The expansive operational diagram of neural nets, I would suggest, ascribes subject positions associated with it. How does that happen? Some familiar diagrammatic operations immediately appear in almost any actual example of a neural net. In the

9 Recent work on deep belief networks replaces the sigmoid function with other nonlinear functions that subtly alter the way layers of neural nets relate to each other. See Glorot and Bengio (2010) for an account of changing training practices in neural nets.

Listing 8.1
Neural net for `titanic` dataset

```
library(neuralnet)

titanic = read.csv("data/titanic3.csv", stringsAsFactors = FALSE)

titanic_transformed = as.data.frame(model.matrix(~survived + age +
    ↪ pclass + fare + sibsp + sex + parch + embarked, titanic))

train_index = sample.int(nrow(titanic)/2)

titanic_train = titanic_transformed[train_index, ]

titanic_net = neuralnet(survived ~ age + pclass + fare + sexmale +
    ↪ sibsp + parch + embarkedC + embarkedQ + embarkedS, data =
    ↪ titanic_train, err.fct = "ce",

    linear.output = FALSE, rep = 1, hidden = 3, stepmax = 1e+05)

titanic_test = titanic_transformed[-train_index, ]

test_error = round(sum(0.5 < compute(titanic_net, titanic_test[, -c
    ↪ (1, 2)])$net.result)/sum(titanic_test$survived), 2)
```

code vignette shown below, the dataset is a spreadsheet of information about passengers of the Titanic. The `titanic` dataset, like `iris` or `boston`, is often used in contemporary machine learning pedagogy. It is, for instance, the main training dataset used by kaggle.com, an online machine learning competition site I will discuss below. The first few lines of the `R` code load the dataset and transform it into vector space. For instance, variables such as `sex`, which take values such as `male` and `female`, become vectors of `1` and `0` in a new variable `sexmale`.

The line of the code that constructs a neural net using the `neuralnet` library (Fritsch and Guenther 2012) closely resembles the lines of code used to construct linear models for the `prostate` data (see chapter 3). The `R` formula for the neural net looks similar to other machine learners such as logistic regression. It models whether someone `survived` the wreck of the Titanic in terms of their age, class of fare (`pclass`), sex, number of siblings/spouse (`sibsp`), number of parents/children (`parch`), and port of departure:

```
survived ~ age + pclass + fare + sexmale + sibsp + parch +
embarkedC + embarkedQ + embarkedS
```

The R model formula expresses the response or target variable `survived` as a combination of other variables. In this case, the plus sign + indicates that the combination is linear or additive. If this model formula looks so similar to other machine learning techniques we have been discussing, then what do neural networks add? Why did and do so many machine learners turn to them?

Perhaps the only distinctive feature of the code listing 8.1 appears in the expression `hidden = 3`. This architectural feature does not appear in the model formula in the R code but does, as we have already seen, operate in the lines of equation 8.1. These hidden units are key to neural net because they construct the derived features that the model learns from the input data X.

How many units are hidden and in what topology matters less than the existence of an operation that allows their topology to be configured in an encounter with data. The novelty of this operation was central to research into neural nets. Rumelhart, Winton, and Williams announce the algorithm in a letter to *Nature* in 1986 titled "Learning representations by back-propagating errors":

> We describe a new learning procedure, back-propagation, for networks of neurone-like units. The procedure repeatedly adjusts the weights of the connections in the network so as to minimize a measure of the difference between the actual output vector of the net and the desired output vector. As a result of the weight adjustments, internal "hidden" units which are not part of the input or output come to represent important features of the task domain, and the regularities in the task are captured by the interactions of these units. The ability to create useful new features distinguishes back-propagation from earlier, simpler methods such as the perceptron-convergence procedure (Rumelhart, Hinton, and Williams 1986, 533).

Again, despite the persistent reference to biology, the description of the "new learning procedure" sounds more like machine learning. There is talk of minimizing a measure of difference between actual and desire output vectors (optimizing through a cost or loss function), as well as mention of "features" and "weights" (usually a synonym for model parameters: "the neural network model has unknown parameters, often called weights, and we seek values for them that make the model fit the training data well" (Hastie, Tibshirani, and Friedman 2009, 395)). The novelty, however, consists of the "hidden" units whose interactions "represent important features." In other words, the flat additive combination of features expressed in the R model formula above does not convey the interactions of these units. (As Hastie and co-authors put it, "the units in the middle of the network, computing the derived features Z_m, are called hidden units because the values Z_m are not directly observed" (Hastie, Tibshirani, and Friedman 2009, 393).) These units can only viably interact in the neural nets because the back-propagation algorithm offers a way to create "useful, new features" from the data. But because they interact through back-propagation, the hidden units capture regularities in the task domain and thereby do what counts as cognition in the connectionist philosophies associated with neural nets (see the PDP group's work; McClelland and Rumelhart 1986).

The hidden layers lend a network form to machine learning. The final diagrammatic form in which neural nets appear is the network. Network graphs already appeared in Rosenblatt's perceptron work (Rosenblatt 1958), but they ramify tremendously in the aftermath of back-propagation. Almost every book and article relating to neural net presents some version of the diagram shown in figure 8.3.

Although the network topology of the model appears in many more complicated forms, it diagrams several operations. First, it presents a surface—the input layer—that indexes something in the world. The input layer, shown as X in the algebraic diagram of equation 8.1, suggests receptive or recording surface (e.g., a camera). Early neural net papers on the handwritten digital recognition problem sometimes describe cameras mounted above tables focused on images (LeCun et al. 1989). Second, it presents an output layer that can contain single or multiple nodes, the k of equation 8.1. In the `titanic` examples, a single target node appears (survived or not). In the `MNIST` handwritten digit recognition models, there are usually 10 output nodes, one for each of the digits 0...9. Third, the network diagram exhibits ordered forms of movement. Data and calculation propagate from left to right or vice versa. (Sometimes the networks are rotated and the flow is vertical but still bidirectional.) Bidirectional hierarchical movement is key to the back-propagation algorithm in feed-forward and more complicated recurrent and convolutional neural nets. Fourth, it renders visible in principle the vital hidden nodes. Without the hidden nodes, neural nets collapse into linear models. With the hidden nodes, the Z_m of the equations 8.1, neural nets, like some other machine learners we have discussed such as support vector machines, effectively expand the vector space by constructing new dimensions in it. The derived features or "learned representations" (to use the language of Rumelhart, Hinton, and Williams 1986) can expand indefinitely, according to different network topologies.

The subjects of a hidden operation

How do the diagrammatic forms of the basic model equations, the network diagram, and the operational code comprising the privileged machine at work recognizing handwritten digits or classifying the passengers on the Titanic assign subject positions?

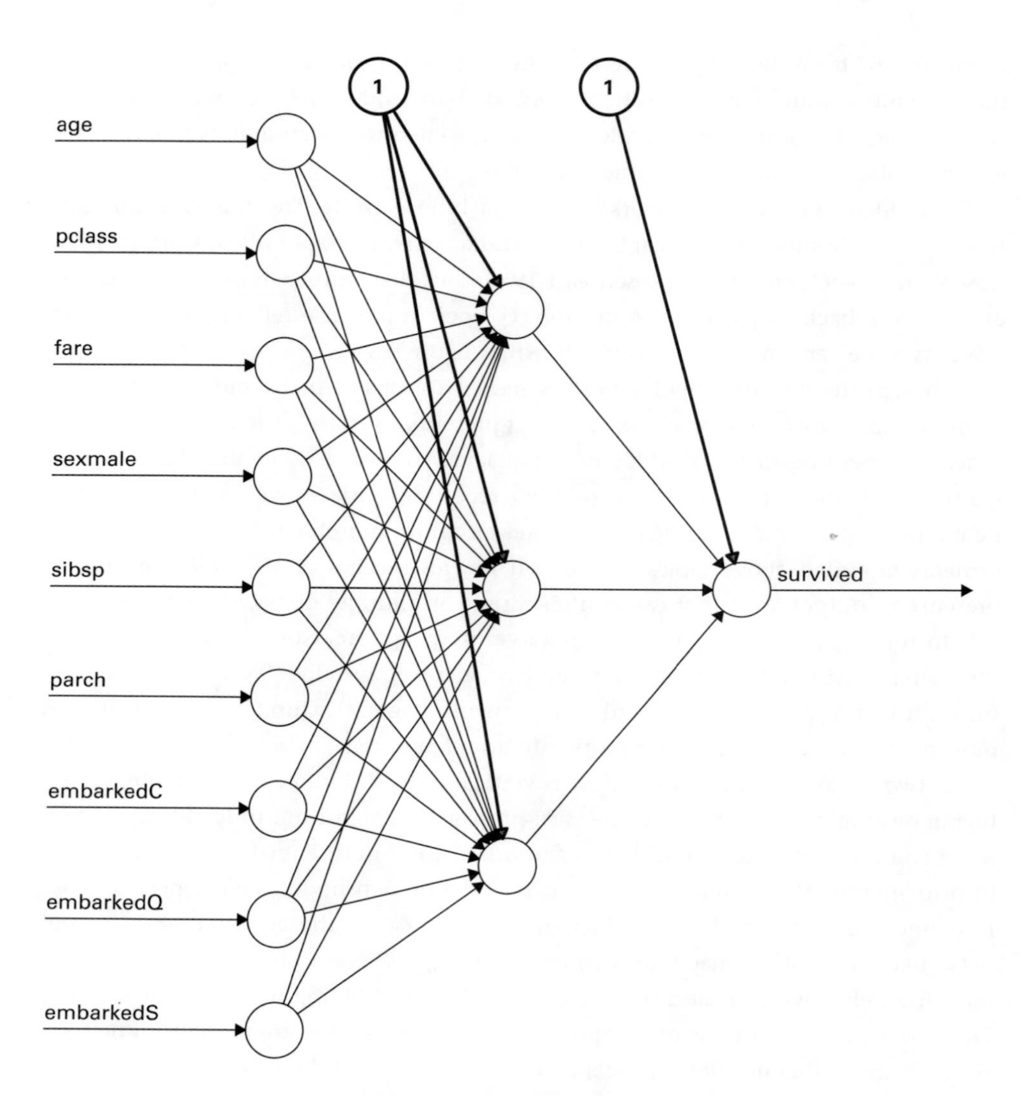

Figure 8.3
Neural network topology for 3-hidden unit "titanic" data.

How would we describe the figure of human machine learner in this setting? Is the human machine learner like Vlad, the former Eastern European mathematician tending the training of a neural net at the heart of the call center knowledge management system, or more like Heather Arthur, the programmer who wrote `kittydar`?

When in *The Archaeology of Knowledge*, Foucault presents the "position of the subject" as an anchor point for the power-laden, knowledge-forming enunciative functions, he does not identify it as a unifying point grounded in interiority, intentionality, or even single speaking position or voice (e.g., that of *the* machine learning expert). On the contrary, "various enunciative modalities manifest his [sic] dispersion" (Foucault 1972, 54). Positions derive from operations that determine statements that become a kind of law for the subject.

As Foucault puts it, using a mathematical example and a conceptual formulation that broadly anticipates accounts of performativity,

> in each case the position of the subject is linked to the existence of an operation that is both determined and present; in each case, the subject of the statement is also the subject of the operation (he who establishes the definition of a straight line is also he who states it; he who posits the existence of a finite series is also, and at the same time, he who states it); and in each case, the subject links, by means of this operation and the statement in which it is embodied, his future statements and operations (as an enunciating subject, he accepts this statement as his own law) (Foucault 1972, 94–95).

Foucault's examples here include subjects who say things such as, "I call straight any series of points that ...": just such statements operate in machine learning. The *operation* is crucial because it connects many different practices and techniques (function finding, optimization, transformation of data into the vector space, mobilization of probability distributions as a kind of rule of existence for learnable situations, etc.) that accompany, ornament, strengthen, and diagram the statement.

Foucault posits, therefore, a subject-positioning circularity between the operation and accompanying statement: the subject of the statement is also the subject of the operation. The provisional coincidence of operation and statement defines a subject position with some agency and subjects machine learners to future operations. The process might be formalized for any machine learner as follows: the diagrammatic operations of the machine learner support the production of statements, and these operations become a way of producing future statements to the extent that the subject of the operation is *also* the subject of the statement. The assignation of a subject position occurs in this forward and backward, feed-forwarding and back-propagating movement between operation and statement.

Algorithms that propagate errors

The distinctive feature of neural nets, at least in their ordinary "vanilla" forms, consists of their use of a series or chain of gradient descent operations to minimize errors by adjusting the weights (or parameters) of all the nodes (or linear models) comprising the network. Adjusting the parameters of the nodes in the neural net hardly seems a striking achievement. However, if we look more closely at the way in which the "internal representations" (Rumelhart, Hinton, and Williams 1986, 536) are iteratively constructed in neural nets, something more interesting begins to emerge from the forward and backward movement of this algorithm. Does an algorithm such as back-propagation diagram the slippery coincidence of subject of operation and subject of statement in machine learning?

The subject-positioning zone of slippage between statement and operation appears as error. Although the gap between operation and statement might seem small, there are many slippages and divergences in it. A minor statement such as "We see that Net-5 does the best, having errors of only 1.6%, compared to 13% for the 'vanilla' network Net-2" (Hastie, Tibshirani, and Friedman 2009, 407) bears within it, in its coupling to all the operations comprising Net-5, a set of determinations, sites, and relations for variously positioned subjects. (These might include machine learners, such as Hinton or Le Cun, but also U.S. Postal workers, whose work must have more or less disappeared as automatic mail sorting improved.) In any concrete situation, in relation to any specific machine learner, the diagrammatic operations and statements will position subjects in specific ways. There is no simple referent here, no simple object gripped or seen by a knowing or controlling subject, because on this account, the operations and statements in their dispersions, accumulations, and distributions overflow any simple dyadic relation between a subject-object or human–machine world.

As we have seen in chapter 4, error rates, training error, test error, generalization error, and validation error criss-cross between human and machine learners. Errors render operations as statements. Although not all of these errors figure directly in the algorithms, the learning procedure of most machine learners derives from the way they update model parameters in the light of statements of errors. Every machine learner makes different determinations in relation to model parameters and errors. We have already seen something of the forward movement that runs between the input and output layers with its classificatory statements. Equation 8.1 implies data moving a succession of layers and their nodes. Conversely, the back-propagating phase of a neural net moves from output toward input layer updating weights of various nodes in the light of differences between predicted and known outputs.

$$\beta_{km}^{(r+1)} = \beta_{km}^{(r)} - \gamma_r \sum_{i=1}^{N} \frac{\partial R_i}{\partial \beta_{km}^{(r)}}$$

$$\alpha_{ml}^{(r+1)} = \alpha_{km}^{(r)} - \gamma_r \sum_{i=1}^{N} \frac{\partial R_i}{\partial \alpha_{km}^{(r)}} \qquad (8.3)$$

(Hastie, Tibshirani, and Friedman 2009, 396)

Many different parameters figure in the back-propagation functions shown in equation 8.3. They include measures of error (R), values of the weights or parameters in various layers of the models (β, α), variables that count the number of iterations the model has performed ($r, r+1$), and the functional operators such as summation ($\sum$) and partial differentiation (∂). The usual indexical relations to vector space appear in N, the number of rows or observations, as well as K, the number of outputs, and M, the number of nodes in the hidden layer. Partial derivatives express the sensitivity of errors with respect to the weights or parameters of the nodes. In the densely iconic and indexical diagram of equation 8.3, the interweaving of the subscripts in the two lines shows how values of the model parameters of the first two lines of equation 8.1 update as the model is trained on the input data. The two lines of equation 8.3 specify how first the values of the parameters β_{km} of the K nodes of the output layer should be altered in the light of the difference between the actual and expected output values and then how the weights α_{km} of the M nodes of the hidden layers should be adjusted. Once these are adjusted, the forward movement defined by equation 8.1 begins again. In adjusting weights in the layers, back-propagation always starts at the outputs and travels back into the net toward the input layer at the bottom (or left-hand side in diagram 8.3).

"It is as if the error propagates from the output y back to the inputs and hence the name *back-propagation* was coined," writes Alpaydin (Alpaydin 2010, 250). As in any gradient descent operation (see chapter 4), a rate parameter (here γ) regulates the speed of movement. If γ is too large, then the gradient descent might jump over a valley that contains the absolute minimum error; if γ is too small, then the descent is too slow for fast machine learning. In some versions of neural net, the value of γ changes each at iteration r of the model.[10]

10 If back-propagation was formulated in the 1980s (and indeed was already known in 1960), what do we learn from its current reiterations? Given the effort that went into crafting neural nets to recognize handwritten digits during the 1980s and 1990s, what does the revival of neural nets suggest about machine learning as a feed-forward/back-propagation operation? From the

Competitions as examination

More or less directly, the observation of error rates converging toward minimum values assigns a subject position the role of controlling hyperparameters such as γ, the learning rate. This seems a drastic curtailment of agency. The related feed-forward and back-propagation of errors focuses the machine learner subject, the "wonderful people" of Hilary Mason's exhortation to developers, on error. At almost every step of its development as a field and in almost every aspect of its operation, competitions to observe and rank error rates bring human and machine learners together. In competitions, errors are not purely epistemic. They circulate within a wider economy of competitive optimization that connect them to power, value, and agency dynamics. The learning of machine learning takes place in examinations that rank both human and nonhuman machine learners according to error rates. What can we learn from such "error-prone" competitions about subject positions in machine learning?

Backward and forward movement between human and machine learners characterizes competitions run by Kaggle. Kaggle organizationally implements a parallel architecture machine learning process by back-propagating errors to hidden nodes embodied in individual competitors who, in principle at least, are not connected to each other but only to the layers and nodes of Kaggle as a platform. In comparison to the research-oriented machine learning competitions such as the annual KDD Cup (KDD 2013) run by the Association of Computing Machinery (ACM) Special Interest Group on Knowledge Discovery and Data Mining, the NIPS (Neural Information Processing Systems) Challenges, the ImageNet Large Scale Visual Recognition Challenge (ILSVRC 2014), or the International Conference on Machine Learning (ICML)[http://machinelearning.org/icml.html], the Kaggle competitions attract a wide range of academic, industry, commercial, and individual entries.

---

early publications (Rumelhart, Hinton, and Williams 1985) on, the layered composition of the model has been linked to architectural considerations. As Hastie and co-authors write:

> The advantages of back-propagation are its simple, local nature. In the back propagation algorithm, each hidden unit passes and receives information only to and from units that share a connection. Hence it can be implemented efficiently on a parallel architecture computer (Hastie, Tibshirani, and Friedman 2009, 397).

These practical considerations have different significance in different settings. Some of the current iterations of neural nets in deep learning rely on massively parallel computing architectures (e.g., Andrew Ng's GoogleX YouTube video project). Yet the information sharing that happens during back-propagation might also encompass the human others of neural nets. The efficient parallel implementation in computing architecture affects, I would suggest, human and nonhuman machine learners in different ways.

Competitors enter these competitions for various reasons, not the least of which is their employment prospects or the promotion of their machine learning products (e.g., the winner of a major competition, the Heritage Health Fund Prize, in 2012 used that prize to promote the data mining software made by his company (Tiberius); an entrant in the Hewlett "Automated Essay Competition," again in 2012, included Pacific Metrics, a U.S. company whose automated essay scoring products were already in use in U.S. schools; although Pacific Metrics did not win the competition, it acquired the winning machine learner and incorporated it into its products (Kaggle 2012)). Kaggle.com is effectively a recruitment agency for machine learners (Kaggle 2015c). For instance, several competitions sponsored by Facebook have positions as data scientists at Facebook as the prize:

> Ever wonder what it's like to work at Facebook? Facebook and Kaggle are launching an Engineering competition for 2015. Trail blaze your way to the top of the leader board to earn an opportunity at interviewing for a role as a software engineer, working on world class Machine Learning problems (Kaggle 2015b).

Although most employment agencies rely on CVs (curriculum vitae), Kaggle operates more like feed-forward and back-propagation between multiple competitions as a way of optimizing its ranking of machine learners.

The competition organizers list three injunctions: download (the data), build (a model), and submit (an entry or many entries to the competition). Leader-boards and individual rankings within the Kaggle's "world's largest community of data scientists" (Kaggle 2015a)[11] allow clients—corporations mostly—to "harness the 'cognitive surplus' " (Kaggle 2015e). Figure 8.4 also shows some of the typical diversity of the several hundred machine learning competitions that Kaggle has staged since 2011: diabetic retinopathy and west Nile virus prediction competition appear next to search results relevance or context ad clicks competitions. As we have seen frequently, accumulation, aggregation, and proximity, whether accidental or constructed, between disparate entities suggest that machine learners possess epistemic mobility not readily available to the domain experts (in diabetes, virology, information retrieval, or search engine optimization).

Like a neural network with many layers and nodes, competitions subject competitors (several hundred thousand in Kaggle) to ranking and indeed prediction based on the generalization error of the models that they submit to the competition. The leaderboard, which displays current rankings of competitors in a given competition, is the visual form of this error-based ranking:

11 At the time of writing, Kaggle claims around 320,000 competitors.

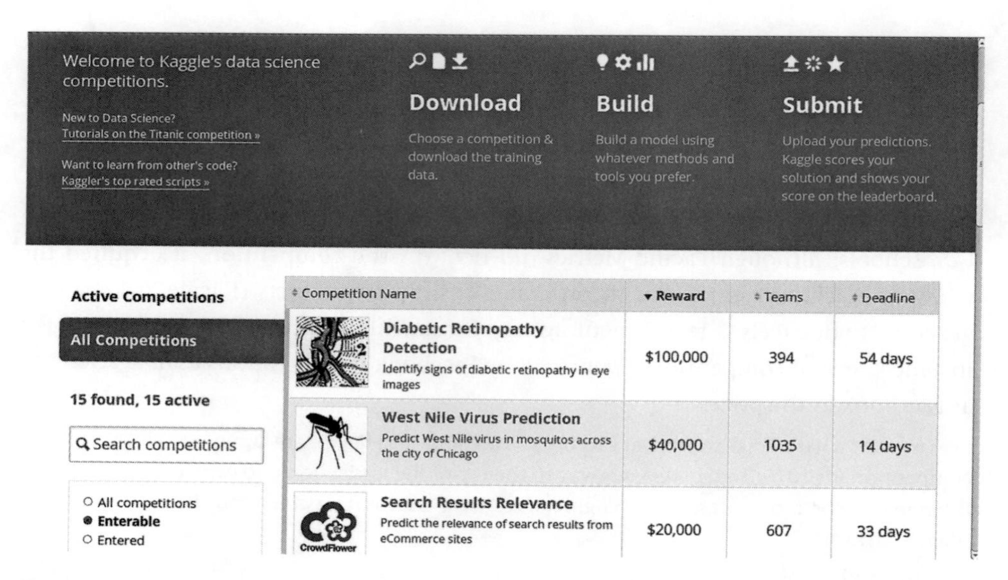

Figure 8.4
Kaggle data science competitions.

> The leaderboard is a central fixture of the Kaggle experience. It provides context to the incredible work accomplished by the Kaggle data science community. To a competitor, the leaderboard is a dynamic, living, action-filled battle. Tactics come to life. Individuals leapfrog over each other. Teams merge and blend submissions. Some submit early and often, attempting to build up insurmountable leads. Others bide time, waiting to pounce minutes before the buzzer with their finest of forests. We see the joys of regularization and the agony of overfitting.... It's thousands of hours of collective human toil (Kaggle 2015e).

The dynamics of ranking and the experience of being ranked here arise from a fairly simple mechanism. Entrants in a given competition download two datasets: a training dataset that includes labels for all the response variables, and a test dataset that does not include the labels. In principle, competitors construct machine learners using the training dataset, implement a machine learner to predict labels for the test dataset, and then upload the predicted labels to Kaggle as a submission to the competition. The Kaggle platform then calculates a ranking based on the generalization error in the test labels.[12] Competitors optimize their entries against each other, but

12 Kaggle has the actual labels for the test dataset. Entrants monitor the leaderboard and attempt to improve their rankings by making new submissions with improved or altered models. The many entries that participants sometimes submit to the competitions suggest that rankings and their visibility operate like the loss functions that optimize the fit of a subject position to an operational formation.

the competition overall functions as a kind of general optimization process in which many hidden nodes adjust their treatment to the training data as scores and rankings propagate through via the leaderboard system. The stylized injunction to download-model-submit many times effectively creates an algorithmic process in which many hidden nodes operate in parallel to produce predictions.

Superimposing power and knowledge

It would be possible to explore in much greater ethnographic depth the practices of Kaggle competitors, the spectrum of participants (ranging from undergraduate student teams through to retired scientists, from hedge fund financial analysts to physicists), and the ways in which the topics of competitions relate to different scientific, governmental, and commercial problems. Here I am interested mainly in the form of the competition as a test or an examination centered on errors. The competitions take the form of examinations that set a problem, define some limits or constraints on its solution, and create a space that qualifies, ranks, and displays the work of individuals or groups according to rates of generalization error (the error that arises when a machine learner encounters new data).

Machine learning competitions instance practices of examination that Foucault described in *Discipline and Punish*:

> The examination combines the techniques of an observing hierarchy and those of a normalizing judgement.... It establishes over individuals a visibility through which one differentiates them and judges them. That is why, in all the mechanisms of discipline, the examination is highly ritualized. In it are combined the ceremony of power and the form of the experiment, the deployment of force and the establishment of truth. At the heart of the procedures of discipline, it manifests the subjection of those who are perceived as objects and the objectification of those who are subjected. The superimposition of the power relations and knowledge relations assumes in the examination all its visible brilliance (Foucault 1977, 183–185).

The disciplinary form of the examination of errors links statements and operations. Examinations combine ceremony, ritual, experiment, force, and truth in subject and object positioning operations. The consolidation of machine learning as a data practice today in competitions occurs via a pervasive practice of examining and testing. The forms of visibility created by competitions individualize and normalize machine learners (often by proper names) and optimize extractions of force, time, propensities, and aptitudes.

In many Kaggle competitions (some titles are shown in table 8.1), winning entries come from machine learners working together. In the National Data Science Bowl

Table 8.1
The highest prize money machine learning competitions on Kaggle

Competition	Reward amount	Domain	Type of data
Heritage Health Prize Identify patients who will be admitted to a hospital within the next year using historical claims data Enter by 06 59 59 UTC Oct 4 2012	500000	health	measurements
GE Flight Quest Think you can change the future of flight	250000	transport	events
Flight Quest 2 Flight Optimization Milestone Phase Optimize flight routes based on current weather and traffic	250000	traffic	events
Flight Quest 2 Flight Optimization Main Phase Optimize flight routes based on current weather and traffic	220000	traffic	measurements
Flight Quest 2 Flight Optimization Final Phase Final Phase of Flight Quest 2	220000	traffic	measurements
National Data Science Bowl Predict ocean health one plankton at a time	175000	science	images
The Hewlett Foundation Automated Essay Scoring Develop an automated scoring algorithm for student written essays	100000	education	texts
The Hewlett Foundation Short Answer Scoring Develop a scoring algorithm for student written short answer responses	100000	education	texts
GE Hospital Quest Think it s possible to make hospital visits hassle free GE does	100000	health	actions
Diabetic Retinopathy Detection Identify signs of diabetic retinopathy in eye images	100000	medicine	images
Allstate Purchase Prediction Challenge Predict a purchased policy based on transaction history	50000	retail	transaction
Merck Molecular Activity Challenge Help develop safe and effective medicines by predicting molecular activity	40000	medicine	measurements
West Nile Virus Prediction Predict West Nile virus in mosquitos across the city of Chicago	40000	science	measurements
Acquire Valued Shoppers Challenge Predict which shoppers will become repeat buyers	30000	retail	transaction
Driver Telematics Analysis Use telematic data to identify a driver signature	30000	traffic	measurements
Restaurant Revenue Prediction Predict annual restaurant sales based on objective measurements	30000	retail	attributes
Caterpillar Tube Pricing Model quoted prices for industrial tube assemblies	30000	retail	attributes
GigaOM WordPress Challenge Splunk Innovation Prospect Predict which blog posts someone will like	25000	social_media	texts
U S Census Return Rate Challenge Predict census mail return rates	25000	government	actions
Belkin Energy Disaggregation Competition Disaggregate household energy consumption into individual appliances	25000	energy	actions

competition of 2015, competitors were asked to classify images of more than 100 species of plankton. The winning team comprised seven graduate and postdoctoral researchers from Ghent University, Belgium. In a jointly written blog account of their winning entry, team Deep Sea describe something of the construction of the deep learning models they built. These were convolutional neural nets, neural nets in which elements of the network only "look" at overlapping tiles of the input images:

> We started with a fairly shallow models by modern standards (~6 layers) and gradually added more layers when we noticed it improved performance (it usually did). Near the end of the competition, we were training models with up to 16 layers. The challenge, as always, was balancing improved performance with increased overfitting (Dieleman 2015).

Like many of the entrants in image-based classification competitions, such as the ImageNet Large Scale Visual Recognition Challenge (ILSVRC 2014), Deep Sea built their machine learner in several stages, first deriving features from the data by creating various layers that looked for common features across various scales, rotations, and other transformations of the plankton images, and then adding neural net layers to classify those derived features using the labels supplied in the training set. In this respect, and in almost perfect synchrony with the deep learning teams at Google, Facebook, and many other places, Deep Sea combined supervised and unsupervised learning techniques. The lower convolutional layers that process the images are strictly speaker unsupervised because they make no use of the known labels or categories of the plankton; the upper layers are supervised because they make use of the labels in the normal back-propagation process of neural net training.

Compared with the plain or "vanilla" neural nets discussed above, deep belief networks involve many more parameters, stages of observation and modeling, configuration of hardware and infrastructural arrangements, and comparison of results. Deep Sea describe the architecture of one of their more successful models:

> It has 13 layers with parameters (10 convolutional, 3 fully connected) and 4 spatial pooling layers. The input shape is (32, 1, 95, 95), in bc01 order (batch size, number of channels, height, width). The output shape is (32, 121). For a given input, the network outputs 121 probabilities that sum to 1, one for each class.

They go on to describe the different layers—cyclic slice, convolutional, spatial pooling—that derive features from the data or augmenting it (by examining overlapping tiles, by rotating or scaling the images, so that any given image is "seen" in a number of different ways, and the model learns to detect these variations). The combination of diverse layers in a stratified model introduces a range of learners into the operation, just as Kaggle networks many machine learners through its competitions.

A massive parallel computation allows deep learning. Infrastructure and cognition entwine heavily here because the possibility of training large many-layered neural nets depends heavily on vectorized transformations of image data. Probably few other competitors in this competition would have had access to the Tesla K40 or NVIDIA GTX 980 Superclocked GPU cards on which Deep Sea relied.[13] Even with that intensive computational resource, their models required "between 24 and 48 hours to reach convergence." They constructed around 300 models. Because of the plethora of models with different architectures and parameters, "we had to select how many and which models to use in the final blend" (Dieleman 2015). As is often the case, competition engenders populations of machine learners whose aggregate tendencies model optimum performance.[14] The DeepSea team might epitomize machine learning subject positions. Like the "wonderful people" described by Hilary Mason, they bring together infrastructure, engineering, mathematics/statistics, and some knowledge of human behavior (although the knowledge of human behavior in this case might have more to do with what other Kaggle competitors might be doing, as well as an awareness of cutting-edge research leaders in image-recognition techniques).

Ranked subject positions

DeepSea built models that classify images of more than a hundred kinds of plankton with few errors. In driving down error rates more than the hundreds of other competitors, they occupy a privileged subject position at the conjunction of operation and the statements in machine learning. Machine learners such as deep belief nets adjust and align subject positions through their many convolutional layers. They supplant, for instance, the skilled configuration of feature engineering that characterized work on decision trees, linear regressions, support vector machines, and predecessor neural nets (and appears as a key element in figure 8.1). Similarly, they absorb the

13 As another competitor in the National Data Science Bowl mentions:

> One example is here the Kaggle plankton detection competition. At first I thought about entering the competition as I might have a huge advantage through my 4 GPU system. I reasoned I might be able to train a very large convolutional net in a very short time—one thing that others cannot do because they lack the hardware (Dettmers 2015).

Hardware parallelism and vectorization, at least in the area of deep learning, seem to matter more than the ability to test, examine, observe, or invest new model configurations.

14 On the command line, `git clone https://github.com/benanne/kaggle-ndsb` makes a copy of the model code. The code in that github repository gives some idea of the mosaic of techniques, configurations, variations, and tests undertaken by DeepSea.

professional skills of Go players in training models that win against the best human players (Silver et al. 2016). The subject position of a machine learner occupies a zone of diagrammatic slippage between statements and operations.

The various subject positions that might speak of, observe, question, or decide about machine learning are neither unified nor fixed. As the models grow, for instance, they test the capacity of human machine learners to understand how models transform data. Perhaps more profoundly, the growth of neural nets exhibits the deeply competitive imperative that imbues much machine learning practice, and in many way machine learning practice. This competition is not always explicit or overt, but it almost transpires in the form of a test or examination.

Neural nets reiteratively draw human–machine learning differences. Their own ups and downs, the merging and blending of statistics, computer science, and cognitive science they afford, and their potential to drive down error or learn features from data given enough data derive less from some exotic mathematical abstraction or encompassing algorithm and more from competitively accumulated layers and connections between units of modeling. The shuttling movement of the central algorithm—feed-forward and back-propagation—is instructive. Because it propagates errors to all elements of the network, and every element in the network adjusts its weights in trying to minimize error, layers can multiply on many scales. The predictive power of the model derives from the networked collective of elementary machine learners driven to optimize their error rates. So, too, the competitive examinations that today generalize machine learning as a data practice predicate the ongoing potential of hidden layers—machine learners—to collectively learn from their rankings in tests of error.

As it disperses subject positions, the back-propagation of errors or optimization also animates optimism about machine learning.[15] Machine learning hovers in potentiality because neural nets and their kin assimilate and adjust their weights in response to changes in infrastructures and in the generalization of operations to newly adjacent

15 The cultural theorist Lauren Berlant describes optimism as an "operation":

> The surrender to the return to the scene where the object hovers in its potentialities is the operation of optimism as an affective form (Berlant 2007, 20).

Berlant's complicated formulation brings together surrender, return, scene, object, potentialities, and affective form. These movements, places, and things might be understood as purely psychic or semiotic processes. But optimism as an asignifying diagrammatic operation also plays out across the manifold surfaces of algorithms, datasets, models, platforms, and ranking systems associated with machine learning as a competitive examination.

domains. Machine learners generate optimism through and about optimization, an optimization that is predictive, prospective, and anticipatory. But this adjusting of weights carried out through the propagation of errors is also inherently a ranking or an examination.

Human and machine learner differences can be redrawn in two different directions. In one direction, machine learning operations assign a subject position focused on error rates. Vlad in his corner observing the neural nets occupied such a position. In the other direction, the subjects who operate the neural net to fit a model find themselves deeply caught up in a network of machine learners connected into parallel and layered architectures and operations. This feeding-forward, however, is regularized or narrowed down through examination and error, through back-propagation on various scales that ranks and filters machine learners according to their error rates. In this direction, the practice of training and testing generalization error that has long guided the supervision of machine learners becomes a mechanism for adjusting subject positions of human machine learners. Some will be wonderful people, some will remain remote like Vlad, and some will learn to optimistically change their ranking.

9 Conclusion: Out of the Data

```
These diagrams of the diagrammatic domains, they kernel together in
localization.

In this contrusion of major forms of invention in natures in machine
learning techniques, inter-places, leveraged in and distributed.
```

The two sentences above are the products of a generative model trained on the raw text of this book. Without any model of syntax, any dictionary of words or terms, relying purely on character sequences as probability distributions, the neural network that sampled these sentences out of its own unsupervised model of the book vectorised as data was primed with starting text of "`If`." "Diagrams of the diagrammatic domains," kerneling together in localization, a "contrusion" of major forms of invention in natures, in machine learning techniques, leveraged in and distributed in interplaces: all of that has been put quite well by the generative model, a two-layer "long short term memory" recurrent neural net (Karpathy 2016).

I began with a relatively open question: if machine learning transforms the production of knowledge, then might the practice of critical thought itself change, whether in its empirical or theoretical orientations? Could the "experimentation of concepts" (Stengers 2000, 153) work with machine learning? My answer is provisionally affirmative. A critical machine learner would learn machine learning to diagram a diagrammatic domain. For such a machine learner, predictions would figure less as statements that rank, order, and classify than as a technology of critical experimentation, a means of effecting a certain number of transformative operations on one's own conduct, thinking and ways of being amid the determinations of contemporary reality. It would function as a mode of experimentation on operations.

230,000 machine learners

For at least 230,000 human machine learners—the population of unique authors listed in the corpus of machine learning research literature I have been drawing on—a new kind of operational formation jells in machine learning. People and things, knowledge and power, combine in novel forms to generate statements. Understanding the distribution and production of elements that make up this emerging common space of decision, classification, prediction, and anticipation directly matters to contemporary critical thought in its engagement with power, production, conduct, communication, ways of being and thinking, materiality, and experience.

Let us take 146,000 scientific articles, publications, and books as statements concerning operations occurring in a variety of sites, modes, and settings connected in the operational formation we are discussing. Statements in operational formations function by reference to the position of a subject (the expert, the engineer, the doctor, the patient, the judge, the teacher, the student) amid an organized or grouped accumulation of devices, settings, and fields (positivity), and with greater or lesser reference to the practices of human–machine interaction: for instance, writing the code that allows the recurrent neural net to build a generative model of this text.

Even if subjects for Foucault do not author statements, the assignment of subject positions in discursive formations always passes through a human subject. In operational formations, subject positions are less distinct, although still highly populated (as the 230,000 authors of these papers suggest). The machine-human mixing in operational formations is highly variable, dynamic, and mutable, sometimes planing through code, sometimes diagrammed in visible forms such as graphs and tables, and often ramifying through infrastructures.

Affective elements have a longstanding connection with computation. Elizabeth Wilson's study, *Affect and Artificial Intelligence* (Wilson 2010), draws on a combination of psychoanalytic, psychological, and archival materials discussing the work of key figures in the early history of artificial intelligence such as Alan Turing on intelligent machinery, Warren McCulloch and Walter Pitts on neural nets, and recent examples of affective computing and robots such as the MIT robot Kismet. Her framing of the psychic nexus with machines such as the perceptron is provocatively operational:

> Sometimes machines are the very means by which we can stay alive psychically, and they can just as readily be a means for affective expansion and amplification as for affective attenuation. This is especially the case of computational machines (Wilson 2010, 30).

Under what conditions do machines and, for present purposes, computational machines become "the very means we can stay alive psychically"? Wilson addresses this question by positing "some kind of intrinsic affinity, some kind of intuitive alliance between the machinic and the affective, between calculation and feeling" (31) and suggesting that the "one of the most important challenges will be to operationalize affectivity in ways that facilitate pathways of introjection between humans and machines" (31). Introjection, the process of bringing the world within self, is, according to psychoanalytic accounts of subjectivity, crucial to the formation of "a stable subject position" (25). Wilson envisages introjection of machine processes as a good, not as a failure or an attenuation of relation to the world.

Although I share Wilson's interest in "affective expansion," I don't see that expansion as unfolding from introjection but rather from an intensification of diagrammatic processes, the act of creating a "concrete being, an intersecting of references" or abstraction (Stengers 2000, 85).

A summary of the argument

I have been experimenting with abstraction in the midst of the accumulation of sites, data, and devices associated with machine learning. Let me resume the argument of the book, an archaeological argument that excavates seven major facets or intersecting planes that belong to the machine learning as an operational formation. Chapter 2 addressed the problem of where amid the mire of data, mathematics, code, infrastructures, scientific, and other knowledge fields, a critical engagement with machine learning might situate itself. I suggested that we should consider the formal, mathematical abstractions, and certain transformations in the production of software associated with machine learning in tandem as diagrammatic processes that organize and assemble human–machine relations. Amid a great accumulation of statements, figures, techniques, constructs, datasets, and code implementations derived from many settings, the task is to map the intersecting references, the diagonal connections, and the transformations and substitutions that weave through machine learning. The positivity of machine learning, its specific forms of accumulation, regularity, and rarety do not attest to the power of algorithms but rather lend liveliness to the field by concentrating expressions from many regions.

Chapter 3 examined the practices of vectorizing data, situating machine learners in an epistopically organized, dimensioned space accommodating an increasing repertoire of transformations operating on vectors. Viewed as another mutation of the tabular grid, vector space invites expansive transformations of data. Machine learning is a practice of working with data to accommodate all differences within an expanding dimensional space, a space in which data are under the strain of smooth surfaces, straight lines, regular curves, and hyperplanes. In terms of both infrastructure and epistemic cultures, the vector space abstracts and concretizes spaces inside data.

What is learning for machine learners? If information and computation can be understood as responding to a crisis in control, then what do machine learners do? Chapter 4 examined how learning institutes experimental relays between operation and observation in optimizing functions that predict and classify. The proliferation of methods and devices in machine learning and the attempts to unify them as "learners" was understood as a result of this entwining of operations and observations. The interplay between operational transformations and observational functions in optimization accounts for much of the "learning" effect in machine learning.

An important and wide-reaching critical strand of work in humanities and social sciences over the last few decades has focused on knowledge in its entanglements with apparatuses of governmentalized power. Populations and other large aggregates have been central objects of concern. They remain so in contemporary operational formations, although under somewhat altered conditions. Having all the data, chapter 5 suggested, is not the principal stake in contemporary data cultures. Instead, the probabilization of both data and machine learners as populations, as distributed probabilities, indicates a different axis along which power-knowledge develops in machine learning.

What happens to differences amid vectorization, learning as optimization, probabilization, and the generalized diagrammatic abstraction of machine learning? Are all differences reduced to quantitative comparisons? Rendered as pattern, chapter 6 explored different treatments of difference in machine learning. Differences bifurcate between infinitesimal graduation and rigid decision boundaries, sometimes blurring or overlapping, and sometimes distributed into inaccessibly high-dimensional inner data spaces. The archaeological task amid the dispersed patterns is to locate differences in kind.

Rather than any new materiality, I have pointed to transformations in referentiality associated with machine learning. From the standpoint of operational archaeology, the materiality of machine learning refers to practices of reuse that stabilize references. Science, by virtue of its experimental inventiveness and truth-authority, cross-validates the referentiality of machine learning. The topic of chapter 7 was a particularly data-intensive contemporary scientific hyperobject, the genome. As a data form, the genome provokes reuse, transcription, and transmission of classifications and predictions. This incites both infrastructural transformations but also new concretizations of the hyperobject (as, for instance, in genome-wide association studies).

Finally, chapter 8 explored the subject position of machine learners. Within operational formations, subject positions arise in gaps between operations and statements concerning operations. The argument here concerned human–machine differences and the dispersion of subject positions through operations that alter those differences. Even among machine learners, subject positions are not fixed or unified. The deep neural networks that beat Go champions in 2015 and 2016 (Silver et al. 2016) or developed hitherto unseen tactics in playing Atari computer games (Mnih et al. 2015) evidence the deeply competitive or test-based administration of this gap.

In-situ hybridization

Alongside the argument concerning abstraction, inclusion, control, multiplicity, differences, materiality, and subject positions, another argument shaped discussion in the preceding chapters, one that effectively underpins the writing. A central problem for critical thought today (and by critical thought I mean post-Foucauldian engagements with the events that constitute us as subjects of what we say, do, and think) concerns how to engage with operational formations. To an even greater extant than the discursive formations that Foucault and many subsequent scholars have analyzed, operational formations in production, communication, and the regulation of conduct become the field in which the work of ethics and politics takes place.

The problem of engagement with operational formations is not so much how to gain control or challenge the asymmetries of access and control that loom so large in them (Facebook can machine learn exponentially more patterns than I can), but to begin to grasp the forms of change that are possible and desirable. Mark Hansen has, for instance, posed the challenge of engaging with data-intensive prediction directly in terms of experience. He writes:

> this imperative enjoins us to use the technologies of data capture, analysis and prediction to create a feed-forward structure capable of marshaling the full productive potentiality of data—its commonality, accessibility, and openness—in order to improve, indeed to improve by *intensifying*, our experience (Hansen 2015, 77).

Treating prediction as more than a means of disciplinary control, and instead as a resource to collectively modulate experience, Hansen's project draws on an extensive engagement with phenomenology and Whitehead's philosophy. The crucial task in his view is creative or inventive: the "feed-forward structure" must marshal "the productive potentiality of data."

One way to do this is broadly aligned with Foucault's emphasis in his later work on care of the self. Technologies of the self "permit individuals to effect a certain number of operations of their bodies and social, thoughts, conduct and ways of being, so as to transform themselves in order to attain a certain state of happiness, purity, wisdom, perfection or even immortality" (Foucault 1997, 225). Could Hansen's feed-forward structure—the term itself referring to the first phase of a neural network machine learner—operate as a technology of the self, not so much focused on improvement or perfection of experience but on the potential to invent new tests of and new relations to pressing realities? For scholars producing critical knowledge in humanities and social science through a variety of textual, empirical, theoretical, and increasingly implicitly or explicitly computational practices, technologies of self offer a concrete path wending a way into domains of production, communication, and governance. Rather than immortality or purity, operations effected on ways of thinking, living, and being might transform oneself in the interests of a limited experience of freedom.

Under what conditions could something like care of the self and technologies of the self have any purchase, relevance, or even toehold in the operational formation of machine learning? Five elements, it seems to me, need to be assembled to think through that conjunction. The recognition of ourselves as subjects of machine learning is a primary archaeological task. Whether in relation to knowledge, communication (in the broadest sense), conduct, or ways of living, this recognition relies on a description of practices associated with differences, multiplicities, materialities, knowledges, and control. Second, as I have endeavored to emphasize in describing machine learning as an operational formation, the liveliness of machine learners should be understood as a localization of power-knowledge relations or a primary field of expressions issuing from many parts (to paraphrase Whitehead). "They kernel together in localization," as my recurrent neural network puts it. Third, while the accumulating plethora of techniques, applications, and sites is unified by neither a master algorithm nor a latent, underlying meaning, it does demonstrate regularities and points of indetermination or slippage. Fourth, understood as a field of the expression of many parts, an operational formation can also be the site of collective individuation. Participating in a collective, individual subjects, far from losing whatever defines their unique or essential identity, gain the chance to individuate, at least in part, the share of preindividual reality that marks the collective within them. Fifth, by participating in a collective, even an operational formation, individuals may transform themselves (to attain certain states or experiences) but also affect the collective itself.

How this might affect the internet filter bubble (Pariser 2011), the "stack to come" (Bratton 2016), digital citizenship (Isin and Ruppert 2015), the character of work (Brynjolfsson and Mcafee 2014), the fabric of experience (Hansen 2015), or what counts as knowledge (Bowker 2014) is hard to say. As an operational formation, machine learning does not determine anything in its operations, even if it connects directly to strategies of power. Foucault writes that "archaeology describes the different spaces of dissension" (Foucault 1972, 152). These spaces of dissension, it seems to me, form a field in which initiatives, individuations, and technologies of the self might articulate a certain number of transformative operations.

Critical operational practice?

Under what conditions would that experimental practice and operation on ways of thinking and saying be divergent rather than convergent? Writing this book, and learning to machine learn in order to write about machine learning, involves participation in a collective, the collective of at least 230,000 scientist-machine learners and the tens of thousands of programmers developing machine learners evident on Github.com. By participating in the collective operational formation, running the risk of being mobilized by existing interests, we might also individuate differently a share of the preindividual reality included within us (Virno 2004, 79). Like Anne-Marie Mol's "praxiography," which seeks to maintain reality multiples in describing practice (Mol 2003, 6), the description of machine learning as data practice intends to sustain the multiple of reality by identifying the practices that make it multiple.

The path I've taken here combines writing (a discursive practice?) and coding (an operational practice). Writing about machine learning is a practice of diagrammatically mapping the reiterative drawing of human–machine relations in code, and in particular in coding that learns from data. Datasets, scientific and engineering publications, textbooks such as *Elements of Statistical Learning*, software libraries and packages, and spectacular demonstrations comprise a whole series of criss-crossings. Even if this is not the path that everyone would or should want to take, for me moving into the data like or as a machine learner perhaps allows writing to become more diagrammatic. "Between the figure and the text we must admit a whole series of criss-crossings," wrote Foucault (Foucault 1972, 66), in defining archaeology as a mode of exploration of knowledges, politics, and ways of being.

Very mundanely, I've read articles and books, downloaded data and software libraries, watched YouTube lectures and presentations, configured and written bits of code and text, made plots and diagrams, and done much configuration work across various platforms (Github.com, linux, Google Compute, `R`, `python`, and `ipython`). Amid all of this data practice (and much practicing), there is no reason to assume that learning machine learning is solely the performance of a conscious subject. When we look at an equation repeatedly, when we comply with the machine learning injunction to "find a useful approximation $(\hat{f})(x)$ to the function $f(x)$ that underlies the predictive relationship between input and output" (Hastie, Tibshirani, and Friedman 2009, 28) by writing code to cross-validate a model, we surrender to "learning" that, however fascinating or surprising, is not that of a conscious human subject but also of human–machine assemblage. To the extent that it is archaeological, operational, diagrammatic writing vibrates around the axis of knowledge/practice, not knowledge/consciousness.

Obstacles to the work of freeing machine learning

As I have emphasized on several occasions, machine learning is an uneasy mixture of massively repeated and familiar forms and something that is not easily understood. On the one hand, the level of imitation, duplications, copying, and reproduction associated with the techniques suggests that a process of remaking the world according to particular forms is in process (e.g., in chapter 5, we saw how Naïve Bayes classifiers are almost demonstrated on spam classification problems). The scientific and engineering literature, with its frequent variations on similar themes, suggests that imitation and copying are very much at the heart of the movements I have been describing. This is nothing new. It would be strange if these techniques were not subject to imitation and emulation. That imitation is predictable. We expect it and can account for it sociologically.[1] Some symptoms of these imitative fluxes can be found in the scientific and engineering literature. As we have seen, work on image and video classification, text and speech, gene interaction prediction, or, above all, predictions of relations or associations between people and things (usually commodities, but not always) is striking in its persevering homogeneity. Moreover, the powerful aspirations evident among large media platforms such as Baidu, Google, and Facebook to relocate machine learners in the field of artificial intelligence amid social media or web page-related data in many ways continues business as usual for computer scientists (Gulcehre 2014).

How would we get any sense of what is not so easily digested and laid out in social practice? Archaeologies of operational formations aim to present some of the necessary elements for that purpose. In the closing pages of *The Archaeology of Knowledge*, Foucault writes:

> the positivities that I have tried to establish must not be understood as a set of determinations imposed from the outside on the thought of individuals, or inhabiting it from the inside, in advance as it were; they constitute rather the set of conditions in accordance with which a practice

1 Accounts that might do this can be found in science and technology studies, particularly in actor-network theory versions, as well as in recent social and cultural theory that, for instance, draws on the work of the 19th-century French sociologist, Gabriele Tarde (Tarde 1902; Borch 2005).

is exercised, in accordance with which that practice gives rise to partially or totally new statements, and in accordance with which it can be modified. These positivities are not so much limitations imposed on the initiative of subjects as the field in which that initiative is articulated (Foucault 1972, 208–209).

Here Foucault refers to the restricted freedom that discursive practices and formations open for us. It is increasingly difficult for science, media, government, and business to think and act outside data. Yet Foucault is quite clear that amid the positivities of knowledge production, knowing the conditions, setting out the rules, and identifying the relations that striate the density and complexity of practice is a precondition to any transformations in practice.

As a data practice, however, machine learning is not entirely predictable. Machine learners, as we have seen, vary too much: they are biased, they overfit, they underfit, and they often fail to generalize. Despite this, they have enormous allure. In the history of automata, automation, and animation, kinetic lures have long exercised fascination, and this may be part of the effect of machine learning. Animating transformations of data (think of the 366 times the logistic regression traverses the `South African Heart Disease` dataset), and then looking at those optimizing animations as "learning" generates operational power dynamics.

Machine learning more broadly attracts infrastructural, technical, professional, semiotic, and financial diagonals—think of the upswing in Google searches for "machine learning" shown in figure 1.1 in chapter 1—that reconfigure it as more thickly transformative, and more "performant." Yet such performant diagrams generate referential effects. Machine learning becomes ontologically potent. As Maurizio Lazzarato writes in *Signs and Machines*, "ontological mutations are always machinic. They are never the simple result of the actions or choices of the 'man' who, leaving the assemblage, removes himself from the non-human, technical, or incorporeal elements that constitute him" (Lazzarato 2014, 83).

New machine learners arise from diagrammatic superimposition of existing practices or procedures. Neural networks are like a massively proliferating nest of perceptrons. Moreover, machine learning techniques often repeat something familiar by different means (think of how `kittydar` treats photographs or how a decision tree is legible but often unfamiliar). The event, then, resides less in either something intrinsic to devices operating as algorithmic models or in something about the domains and places in which the devices operate (biomedicine, state security and intelligence agencies, finance, business, commerce, science, etc.). Perhaps it is a rather more modest event in which the tending of abstractions through estimation, optimization, high-dimensional vectorization, probabilistic mixing of latent and feature variables, and imputation unevenly replaces existing ontological and epistemic norms of verification, objectification, and attribution.

I have been less interested in treating these techniques as the predictable reanimation of alienated reason and more inclined to look for those elements in machine learning that diagrammatically abstract away from structures of representations, subjectification, or indeed implementation associated with platforms, services, and products. Rather than, for instance, interminable implementations of document classifiers, sentiment analyzers, image labeling, handwritten digit recognition, or autonomous navigation, we might instead have a "`contrusion of major forms of invention in natures,...inter-places, leveraged in and distributed.`"

Glossary

$\hat{\beta}$ is a commonly used symbol for the model parameters, weights, or coefficients. Estimating optimum values of β is a preoccupation in machine learning.

$\prod$ is an operator that multiplies together all the terms to the right of the symbol.

$\sum$ is an operator that sums together all the terms to the right of the symbol.

archaeology Michel Foucault defines archaeology as a description that explores the production of statements at the level of knowledge practices (*savoir*). It emphasizes the irregularities and discontinuities in knowledge practices as well as the derivations of operations and functions.

bias of a model refers to its inevitable approximation and misalignment to the actual processes that generated the data.

classifier is a machine learner that assigns instances to classes or categories such as `survive` or `die`, `cat` or `dog`.

cost function is a function that measures the difference between the output of the model (the prediction) and the known values.

cross-validate is an operation that validates a model against a part of the data to gauge how well predictions generalize to fresh or hitherto unseen data. Many rounds of cross-validation may be used in training models when data are limited.

data strain borrows from A.N. Whitehead's notion of strain (Whitehead 1960), which refers to implicit forces or tensions in bodies of data that relate to the feeling of geometrically straight or flat loci.

decision boundary is a boundary or surface constructed in vector space by a machine learning classifier to differentiate or separate and hence classify cases.

deep learning a neural network comprising many layers commonly used for image recognition.

diagram is a form of abstraction concerned with functioning and operations. In Gilles Deleuze's reading of Michel Foucault, diagrams display relations of force and construct models of truth (Deleuze, 1994).

discourse For Michel Foucault, a discourse groups statements generated by an enunciative function.

enunciative function For Michel Foucault, the mapping of statements to themselves, to subject positions, to correlate domains, and their material forms of reuse, replication, and transcription together generate statements. The many predictions, inferences, plots, tabulations, numbers, scores, probabilities, classifications, software libraries, and devices comprise the enunciative function of machine learning.

enunciative modality For Michel Foucault, the sites, forms of observing, describing, teaching, and perceiving associated with statements.

feature Also known in machine learning as variable, measurement, observation, or attribute, a feature occupies one dimension in the vector space inhabited by data. Machine learners often construct new features from and in data.

function Mathematically, a function uniquely maps one set of numbers onto another set of numbers. In machine learning, functions operate diversely, sometimes transforming data to generate feature or vector spaces, sometimes measuring cost or loss for particular models, and sometimes expressing forms such as curves and surfaces that transform data. Across these different usages and domains, the operation of mapping or relation between sets of values such as X and Y can be seen.

generative model uses probability distributions to model the process that generated the data, thus allowing the model to generate or simulate samples from the data.

machine learner refers to humans and machines involved in learning from data together.

operational formation is a variation on Michel Foucault's discursive formation that highlights the collective human–machine regularities of power-knowledge. Although operation and operational fields are implicit to discursive practice, they are somewhat overshadowed by the figures of the document, the utterance, and the proposition in Foucault's account.

partial derivative is an operator from differential calculus that expresses the rate of change of one variable with respect to another.

partial observer in Gilles Deleuze and Félix Guattari's concept of what a mathematical function does in science (Deleuze 1994). A partial derivative is a good example of a partial observer.

perceptron A machine learner developed in the 1950s by Frank Rosenblatt. It is modeled on a neurone that learns to classify the input data or what it "perceives" by varying parameters or weights on the sum of its inputs to produce values of either "1" or "0".

positivity Michel Foucault's term in *Archaeology of Knowledge* to describe the specific forms of accumulation of a group of statements in a discursive formation.

probabilization The process where machine learners themselves are constituted as a population of devices whose distribution and tendencies can be treated statistically.

referential For Michel Foucault, the referential of a statement is not the referent (the facts, things, realities, or beings designated) but the place, condition, field of emergence, or principle of differentiation for the entities named, described, or designated in the statement. The referentials for machine learning include various hyperobjects such as genomes, social media, epidemics, markets, and economies. Such referentials encompass many named entities.

regularization operates on the referentials of machine learning to target subtle, diffuse distributions of difference to classify, estimate, and rank their effects, usually by intensive practices of observation and testing.

statement Michel Foucault's term for the product of an enunciative function that operationally relates a number of elements to a field of objects, establishing subject positions associated with them, and configuring a domain of coordination in which these elements can be invoked, used, and repeated. Statements take many forms, including utterances, graphs, equations, and numbers (Foucault, 82).

variance of a model refers to its dependence on the particular data it is trained on.

vector Three senses of the term are relevant: 1. a vector as an element of vector space; 2. a data structure in programming languages such as R—a one dimensional array of elements; and 3. a feeling in the sense used by A.N. Whitehead to describe the transfer from "there" to "here".

vector space is a hyperspace of indefinite dimensions generated by the projective mapping of data variables or features into distinct coordinate dimensions.

vectorize operations on data that transform vectors of values in aggregate.

Bibliography

Abramowitz, Milton. 1965. *Handbook of Mathematical Functions: With Formulas, Graphs, and Mathematical Tables*. In collaboration with Irene A. Stegun and United States. National Bureau of Standards. New York: Dover Books.

Ackley, David H., Geoffrey E. Hinton, and Terrence J. Sejnowski. 1985. "A Learning Algorithm for Boltzmann Machines." *Cognitive Science* 9 (1): 147–169.

ACM. 2013. "John M. Chambers—Award Winner." Accessed December 12, 2013. http://awards.acm.org/award_winners/chambers_6640862.cfm.

Adams, Vincanne, Michelle Murphy, and Adele E. Clarke. 2009. "Anticipation: Technoscience, Life, Affect, Temporality." *Subjectivity* 28 (1): 246–265.

Adler, Joseph, and Jörg Beyer. 2010. *R in a Nutshell*. Sebastopol, CA: O'Reilly.

Alpaydin, Ethem. 2010. *Introduction to Machine Learning*. Cambridge, MA; London: MIT Press.

Amoore, Louise. 2011. "Data Derivatives on the Emergence of a Security Risk Calculus for Our Times." *Theory, Culture & Society* 28 (6): 24–43.

Anthony, Sebastian. 2012. "Google Compute Engine: For $2 Million/Day, Your Company Can Run the Third Fastest Supercomputer in the World | ExtremeTech." Accessed October 3, 2012. http://www.extremetech.com/extreme/131962-google-compute-engine-for-2-millionday-your-company-can-run-the-third-fastest-supercomputer-in-the-world.

Arendt, Hannah. 1998. *The Human Condition*. Chicago; London: University of Chicago Press.

Aristotle. 1975. Aristotle's Categories and de Interpretatione. Translated by J.L. Ackrill. New York: Oxford University Press.

Arthur, Charles. 2015. "Artificial Intelligence: Don't Fear AI. It's Already on Your Phone—and Useful." Accessed July 9, 2015. http://www.theguardian.com/technology/2015/jun/15/artificial-intelligence-ai-smartphones-machine-learning.

Arthur, Heather. 2012. "harthur/kittydar." Accessed September 16, 2014. https://github.com/harthur/kittydar.

Bailey, N. T. J. 1965. *Probability Methods of Diagnosis Based on Small Samples*. London: HM Stationery Office, London.

Barber, David. 2011. *Bayesian Reasoning and Machine Learning*. Cambridge; New York: Cambridge University Press.

Barocas, Solon, Sophie Hood, and Malte Ziewitz. 2013. *Governing Algorithms: A Provocation Piece*. SSRN SCHOLARLY PAPER ID 2245322. Rochester, NY Social Science Research Network.

BBC. 2012. "Google 'Brain' Machine Spots Cats." BBC News: Technology (June 26).

Beer, David, and Roger Burrows. 2013. "Popular Culture, Digital Archives and the New Social Life of Data." *Theory, Culture & Society* 30 (4): 47–74.

Bellman, Richard. 1961. *Adaptive Control Processes: A Guided Tour*. Vol. 4. Princeton, NJ: Princeton University Press.

Beniger, James R. 1986. *The Control Revolution: Technological and Economic Origins of the Information Society*. Cambridge, MA: Harvard University Press.

Beniger, James R., and Dorothy L. Robyn. 1978. "Quantitative Graphics in Statistics: A Brief History." *The American Statistician* 32, no. 1 (February 1): 1–11.

Berlant, L. 2007. "Nearly Utopian, Nearly Normal: Post-Fordist Affect in La Promesse and Rosetta." *Public Culture* 19 (2): 273.

Bertin, Jacques. 1983. *Semiology of Graphics: Diagrams, Networks, Maps*. Madison: University of Wisconsin Press.

Bishop, Christopher M., et al. 1995. *Neural Networks for Pattern Recognition*. Cambridge; New York: Cambridge University Press.

Bishop, Christopher M. 2006. *Pattern Recognition and Machine Learning*. Vol. 1. New York: Springer.

Blei, David M., and John D. Lafferty. 2007. "A Correlated Topic Model of Science." *The Annals of Applied Statistics*, 17–35. JSTOR: 4537420.

Blei, David M., Andrew Y. Ng, and Michael I. Jordan. 2003. "Latent Dirichlet Allocation." *The Journal of Machine Learning Research* 3: 993–1022.

Bollas, Christopher. 2008. *The Evocative Object World*. London; New York: Routledge.

Borch, Christian. 2005. "Urban Imitations: Tarde's Sociology Revisited." *Theory Culture Society* 22 (3): 81–100.

Bowker, Geoffrey. 2014. "Emerging Configurations of Knowledge Expression." In *Media Technologies: Essays on Communication, Materiality, and Society*, edited by Tarleton Gillespie, Pablo Boczkowski, and Kirsten A. Foot, 99–118. Cambridge, MA: MIT Press.

Boyd, Stephen, and Lieven Vandenberghe. 2004. *Convex Optimization*. Cambridge; New York: Cambridge University Press.

Bratton, Benjamin H. 2016. *The Stack: On Software and Sovereignty*. 1st edition. Cambridge, MA: MIT Press.

Breiman, Leo. 2001a. "Random Forests." *Machine Learning* 45 (1): 5–32.

———. 2001b. "Statistical Modeling: The Two Cultures" (with Comments and a Rejoinder by the Author). *Statistical Science* 16 (3): 199–231.

Breiman, Leo, Jerome Friedman, Richard Olshen, Charles Stone, D. Steinberg, and P. Colla. 1984. *CART: Classification and Regression Trees*. Belmont, CA: Wadsworth.

Brynjolfsson, Erik, and Andrew Mcafee. 2014. *The Second Machine Age—Work, Progress, and Prosperity in a Time of Brilliant Technologies*. New York: W. W. Norton & Company.

Campbell-Kelly, Martin. 2003. *From Airline Reservations to Sonic the Hedgehog: A History of the Software Industry*. Cambridge, MA: MIT Press.

Cassirer, Ernst. 1923. *Substance and Function. Translated by William Curtis Swabey and Marie Curtis Swabey*. Chicago: Open Court Publishing.

Chen, Xi, and Hemant Ishwaran. 2012. "Random Forests for Genomic Data Analysis." *Genomics* 99, no. 6 (june): 323–329.

Cheney-Lippold, John. 2011. "A New Algorithmic Identity Soft Biopolitics and the Modulation of Control." *Theory, Culture & Society* 28 (6): 164–181.

Church, Alonzo. 1936. "A Note on the Entscheidungsproblem." *Journal of Symbolic Logic* 1 (1): 40–41.

———. 1996. *Introduction to Mathematical Logic*. Princeton, NJ: Princeton University Press.

Cleveland, William S., Eric Grosse, and William M. Shyu. 1992. "Local Regression Models." *Statistical Models in S*, 309–376.

CNN. 2011. "40 Under 40: Ones to Watch." Accessed January 22, 2013. http://money.cnn.com/galleries/2011/news/companies/1110/gallery.40_under_40_ ones_to_watch.fortune/.

Coleman, Gabriella. 2012. *Coding Freedom: The Ethics and Aesthetics of Hacking*. Princeton, NJ: Princeton University Press.

Collins, Harry M. 1990. *Artificial Experts: Social Knowledge and Intelligent Machines*. Inside Technology. Cambridge, MA: MIT Press.

Conway, Drew, and John Myles White. 2012. *Machine Learning for Hackers*. Sebastopol, CA: O'Reilly.

Cortes, C., and V. Vapnik. 1995. "Support-Vector Networks." *Machine Learning* 20, no. 3 (September): 273–297.

Couldry, Nick. 2012. *Media, Society, World: Social Theory and Digital Media Practice*. Cambridge; Malden, MA: Polity.

Cover, Thomas, and Peter Hart. 1967. "Nearest Neighbor Pattern Classification." Information Theory, *IEEE Transactions* 13 (1): 21–27.

Cox, Geoff. 2012. *Speaking Code: Coding as Aesthetic and Political Expression*. Cambridge, MA: MIT Press.

Cramer, J. S. 2004. "The Early Origins of the Logit Model." *Studies in History and Philosophy of Science Part C: Studies in History and Philosophy of Biological and Biomedical Sciences* 35 (4): 613–626.

CRAN. 2010. "The Comprehensive R Archive Network." Accessed May 5, 2010. http://www.stats.bris.ac.uk/R/.

Cranor, Lorrie Faith, and Brian A. LaMacchia. 1998. "Spam!" *Communications of the ACM* 41 (8): 74–83.

Dahl, George. 2013. "Deep Learning How I Did It: Merck 1st Place Interview." Accessed June 17, 2013. http://blog.kaggle.com/2012/11/01/deep-learning-how-i-did-it-merck-1st-place-interview/.

Deleuze, Gilles. 1988a. *Bergsonism*. Translated by Hugh Tomlinson and Barbara Habberjam. New York: Zone Books.

———. 1988b. *Foucault*. Translated by Seân Hand. Minneapolis: University of Minnesota Press.

Deleuze, Gilles, and Félix Guattari. 1994. *What Is Philosophy*? Translated by Hugh Tomlinson. European perspectives. New York; Chichester: Columbia University Press.

Dempster, A. P., N. M. Laird, and D. B. Rubin. 1977. "Maximum Likelihood from Incomplete Data via the EM Algorithm." *Journal of the Royal Statistical Society. Series B (Methodological)* 39, no. 1 (January 1) 1–38. JSTOR 2984875.

Derrida, Jacques. 1989. *Edmund Husserl's Origin of Geometry, an Introduction*. Translated by John Leavey. Lincoln: University of Nebraska Press.

Dettmers, Tim. 2015. "Which GPU(s) to Get for Deep Learning: My Experience and Advice for Using GPUs." Accessed July 8, 2015. https://timdettmers.wordpress.com/2014/08/14/which -gpu-for-deep-learning/.

Dieleman, Sander. 2015. "Classifying Plankton with Deep Neural Networks." Accessed July 3, 2015. http://benanne.github.io/2015/03/17/plankton.html.

Domingos, Pedro. 2012. "A Few Useful Things to Know About Machine Learning." *Communications of the ACM* 55 (10): 78–87.

———. 2015. *The Master Algorithm: How the Quest for the Ultimate Learning Machine Will Remake Our World*. New York: Basic Civitas Books.

Doyle, Peter. 1973. "The Use of Automatic Interaction Detector and Similar Search Procedures." *Operational Research Quarterly*, 465–467. JSTOR 10.2307/3008131.

Dreyfus, Hubert L. 1972. *What Computers Can't Do*. New York: Harper & Row.

———. 1992. *What Computers Still Can't Do: A Critique of Artificial Reason*. Cambridge, MA: MIT Press.

Duda, Richard O., Peter E. Hart, and David G. Stork. 2012. *Pattern Classification*. New York; London: John Wiley & Sons.

Durbin, Richard, Sean R. Eddy, Anders Krogh, and Graeme Mitchison. 1998. *Biological Sequence Analysis: Probabilistic Models of Proteins and Nucleic Acids*. 1st edition. Cambridge; New York: Cambridge University Press.

Edwards, Paul N. 1996. *The Closed World: Computers and the Politics of Discourse in Cold War*. Inside Technology. Cambridge, MA; London: MIT Press.

Efron, B. 1979. "Bootstrap Methods: Another Look at the Jackknife." *The Annals of Statistics* 7 (1): 1–26.

Einhorn, Hillel J. 1972. "Alchemy in the Behavioral Sciences." *Public Opinion Quarterly* 36, no. 3 (September 21) 367–378.

Ensmenger, Nathan. 2012. "Is Chess the Drosophila of Artificial Intelligence? A Social History of an Algorithm." *Social Studies of Science* 42, no. 1 (February 1) 5–30.

Fico. 2015. "FICO® Analytic Modeler Decision Tree Professional | FICOTM." Accessed November 1, 2015. http://www.fico.com/en/products/fico-analytic-modeler-decision-tree-professional.

Fisher, R. A. 1938. "The Statistical Utilization of Multiple Measurements." *Annals of Human Genetics* 8 (4): 376–386.

Fisher, Ronald A. 1936. "The Use of Multiple Measurements in Taxonomic Problems." *Annals of Eugenics* 7 (2): 179–188.

Fix, Evelyn, and Joseph L. Hodges. 1951. *Discriminatory Analysis-Nonparametric Discrimination: Consistency Properties*. DTIC Document.

Flach, Peter. 2012. *Machine Learning: The Art and Science of Algorithms That make Sense of Data*. Cambridge: Cambridge University Press.

Foucault, Michel. 1992 [1966]. *The Order of Things: An Archaeology of Human Sciences*. Translated by Allan Sheridan-Smith. London: Routledge.

———. 1972. *The Archaeology of Knowledge and the Discourse on Language*. Translated by Allan Sheridan-Smith. New York: Pantheon Books.

———. 1977. *Discipline and Punish: The Birth of the Prison*. Translated by Allan Sheridan-Smith. New York: Vintage.

———. 1991. *The History of Sexuality*. Translated by Robert Hurley. London: Penguin.

———. 1997. *Ethics: Subjectivity and Truth*. Edited by Paul Rabinow. New York: New Press.

———. 1998. *The Will to Knowledge: The History of Sexuality*. Translated by Robert Hurley. Vol. 1. London: Penguin.

Frey, Carl Benedikt, and Michael Osborne. 2013. *The Future of Employment: How Susceptible are Jobs to Computerisation*? Oxford: Oxford Martin School, Oxford University.

Friedman, Jerome H. 1997. "On Bias, Variance, 0/1—Loss, and the Curse-of-Dimensionality." *Data Mining and Knowledge Discovery* 1 (1): 55–77.

Fritsch, Stefan, and Frauke Guenther. 2012. "Neuralnet: Training of Neural Networks."

Fuller, Matthew, and Andrew Goffey. 2012. *Evil Media*. Cambridge, MA: MIT Press.

Galloway, Alexander. 2014. "The Cybernetic Hypothesis." *differences* 25, no. 1 (January 1): 107–131.

Galloway, Alexander R. 2004. *Protocol: How Control Exists After Decentralization*. Leonardo (Series) (Cambridge, Mass.) Cambridge, MA: MIT Press.

Garling, Caleb. 2015. "Andrew Ng: Why 'Deep Learning' Is a Mandate for Humans, Not Just Machines | WIRED." Accessed July 9, 2015. http://www.wired.com/2015/05/andrew-ng-deep-learning-mandate-humans-not-just-machines/.

Gillespie, Tarleton. 2010. "The Politics of 'Platforms.'" *New Media & Society* 12 (3): 347–364.

———. 2014. "The Relevance of Algorithms." In *Media Technologies: Essays on Communication, Materiality, and Society*, edited by Tarleton Gillespie, Pablo Boczkowski, and Kirsten A. Foot, 167–194. Cambridge, MA: MIT Press.

Gitelman, Lisa, ed. 2013. *"Raw Data" Is an Oxymoron*. Cambridge, MA; London, England: MIT Press.

Glorot, Xavier, and Yoshua Bengio. 2010. "Understanding the Difficulty of Training Deep Feedforward Neural Networks." *International Conference on Artificial Intelligence and Statistics*, 249–256.

Gomes, Lee. 2014. "Machine-Learning Maestro Michael Jordan on the Delusions of Big Data and Other Huge Engineering Efforts—IEEE Spectrum." Accessed March 5, 2015. http://spectrum.ieee.org/robotics/artificial-intelligence/machinelearning-maestro-michael-jordan-on-the-delusions-of-big-data-and-other-huge-engineering-efforts.

Google Inc. 2012. "Behind the Compute Engine Demo at Google I/O 2012 Keynote—Google Compute Engine—Google Developers." Accessed August 13, 2012. https://developers.google.com/compute/io.

Google Inc. 2015. "TensorFlow—an Open Source Software Library for Machine Intelligence." Accessed June 7, 2016. https://www.tensorflow.org/.

Gruber, John. 2004. "Markdown: Syntax." Accessed July 1, 2013. http://daringfireball.net/projects/markdown/.

Guattari, Félix. 1984. *Molecular Revolution: Psychiatry and Politics*. Harmondsworth, Middlesex, England; New York: Penguin.

Guattari, Felix, and Gilles Deleuze. 1988. *A Thousand Plateaus: Capitalism and Schizophrenia*. London: Athlone.

Gulcehre, Caglar. 2014. "Welcome to Deep Learning." Accessed October 24, 2014. http://deeplearning.net/.

Hacking, Ian. 1975. *The Emergence of Probability*. Cambridge; New York: Cambridge University Press.

———. 1990. *The Taming of Chance*. Cambridge; New York: Cambridge University Press.

Hallinan, Blake, and Ted Striphas. 2014. "Recommended for You: The Netflix Prize and the Production of Algorithmic Culture." *New Media & Society* (June 23): 1–21.

Halpern, Orit. 2015. *Beautiful Data*. Durham, NC: Duke University Press.

Hand, D. J., and K. M. Yu. 2001. "Idiot's Bayes—Not So Stupid After All?" *International Statistical Review* 69, no. 3 (December): 385–398.

Hansen, Mark B. N. 2015. *Feed-Forward: On the Future of Twenty-First-Century Media*. Chicago: University of Chicago Press.

Haraway, Donna. 1997. *Modest-Witness@Second-Millennium.FemaleMan-Meets-OncoMouse: Feminism and Technoscience*. New York; London: Routledge.

Hastie, Trevor, Robert Tibshirani, and Jerome H. Friedman. 2001. *The Elements of Statistical Learning: Data Mining, Inference, and Prediction*. 1st edition. New York: Springer.

———. 2009. *The Elements of Statistical Learning: Data Mining, Inference, and Prediction*. 2nd edition. New York: Springer.

Hayles, N. Katherine. 1999. *How We Became Posthuman: Virtual Bodies in Cybernetics, Literature, and*. Chicago; London: University of Chicago Press.

Heis, Jeremy. 2014. "Ernst Cassirer's Substanzbegriff Und Funktionsbegriff." *HOPOS: The Journal of the International Society for the History of Philosophy of Science* 4, no. 2 (September 1): 241–270. JSTOR 10.1086/676959.

Helmreich, Stefan. 2000. *Silicon Second Nature: Culturing Artificial Life in a Digital World*. Berkeley, CA; London: University of California Press.

Henson, Joseph, German Tischler, and Zemin Ning. 2012. "Next-Generation Sequencing and Large Genome Assemblies." *Pharmacogenomics* 13, no. 8 (June): 901–915. pmid: 22676195.

Hey, T., S. Tansley, and K. Tolle, eds. 2009. *The Fourth Paradigm: Data-Intensive Scientific Discovery*. Seattle: Microsoft Research.

Hinton, Geoffrey E. 1989. "Connectionist Learning Procedures." *Artificial Intelligence* 40 (1): 185–234.

Hinton, Geoffrey E., Simon Osindero, and Yee-Whye Teh. 2006. "A Fast Learning Algorithm for Deep Belief Nets." *Neural Computation* 18, no. 7 (July): 1527–1554.

Hinton, Geoffrey E., and Ruslan R. Salakhutdinov. 2006. "Reducing the Dimensionality of Data with Neural Networks." *Science* 313 (5786): 504–507.

Hof, Robert D. 2014. "Chinese Search Giant Baidu Thinks AI Pioneer Andrew Ng Can Help It Challenge Google and Become a Global Power." Accessed May 18, 2015. http://www.technologyreview.com/featuredstory/530016/a-chinese-internet-giant-starts-to-dream/.

Hood, Leroy, and Daniel J. Kevles, eds. 1992. "Biology and Medicine in the Twenty-First Century." In *The Code of Code*, 136–63. Cambridge MA Harvard University Press.

Hothorn, Torsten. 2014. "CRAN Task View: Machine Learning & Statistical Learning." December 18. Accessed September 22, 2015. http://CRAN.R-project.org/view=MachineLearning.

Hothorn, Torsten, Kurt Hornik, and Achim Zeileis. 2006. "Unbiased Recursive Partitioning: A Conditional Inference Framework." *Journal of Computational and Graphical Statistics* 15 (3): 651–674.

IBM. 2014. "IBM's Watson Learns the Language of Science." Accessed April 16, 2015. https://www-03.ibm.com/press/us/en/pressrelease/44697.wss.

ILSVRC. 2014. "ImageNet Large Scale Visual Recognition Competition 2014 (ILSVRC2014)." Accessed July 6, 2015. http://www.image-net.org/challenges/LSVRC/2014/.

Isin, Engin, and Evelyn Ruppert. 2015. *Being Digital Citizens*. 1st edition. London: Rowman & Littlefield International.

Issenberg, Sasha. 2012. "The Definitive Story of How President Obama Mined Voter Data to Win a Second Term | MIT Technology Review." Accessed January 9, 2013. http://www.technologyreview.com/featuredstory/509026/how-obamas-team-used-big-data-to-rally-voters/.

James, Gareth, Daniela Witten, Trevor Hastie, and Robert Tibshirani. 2013. *An Introduction to Statistical Learning*. New York: Springer.

Jockers, Matthew L. 2013. Macroanalysis: Digital Methods and Literary History. Urbana: University of Illinois Press.

Kaggle. 2012. "The Hewlett Foundation: Automated Essay Scoring | Kaggle." Accessed July 1, 2015. https://www.kaggle.com/c/asap-aes.

———. 2015a. "About | Kaggle." Accessed June 3, 2015. https://www.kaggle.com/about.

———. 2015b. "Competitions | Kaggle." Accessed June 3, 2015. https://www.kaggle.com/solutions/competitions.

———. 2015c. "Data Science Jobs Forum | Kaggle." Accessed July 2, 2015. https://www.kaggle.com/jobs.

———. 2015d. "Description—Facebook Recruiting IV: Human or Robot? | Kaggle." Accessed July 2, 2015. https://www.kaggle.com/c/facebook-recruiting-iv-human-or-bot.

———. 2015e. "Description—Leaping Leaderboard Leapfrogs | Kaggle." Accessed June 4, 2015. https://www.kaggle.com/c/leapfrogging-leaderboards.

Karpathy, Andrej. 2016. "karpathy/char-rnn." Accessed June 28, 2016. https://github.com/karpathy/char-rnn.

KDD. 2013. "Call for KDD Cup." Accessed July 23, 2013. http://www.kdd.org/kdd2013/call-for-cup.

Keating, Peter, and Alberto Cambrosio. 2012. "Too Many Numbers: Microarrays in Clinical Cancer Research." *Studies in History and Philosophy of Science Part C: Studies in History and Philosophy of Biological and Biomedical Sciences* 43, no. 1 (March): 37–51.

Khan, Javed, Jun S. Wei, Markus Ringner, Lao H. Saal, Marc Ladanyi, Frank Westermann, Frank Berthold, Manfred Schwab, Cristina R. Antonescu, Carsten Peterson, et al. 2001. "Classification and Diagnostic Prediction of Cancers Using Gene Expression Profiling and Artificial Neural Networks." *Nature Medicine* 7 (6): 673–679.

Kirk, Matthew. 2014. *Thoughtful Machine Learning: A Test-Driven Approach*. 1st edition. Sebastopol, CA: O'Reilly Media.

Kitchin, Rob. 2014. "Big Data, New Epistemologies and Paradigm Shifts." *Big Data & Society* 1 (1): 2053951714528481.

Klimt, Bryan, and Yiming Yang. 2004. "The Enron Corpus: A New Dataset for Email Classification Research." In *Machine Learning: ECML 2004*, 217–226. New York: Springer.

Koller, Daphne. 2012. "What We're Learning from Online Education." Video on TED.com. Accessed June 12, 2016. https://www.ted.com/talks/daphne_koller_what_we_re_learning_from_online_education.

Krzywinski, Martin I., Jacqueline E. Schein, Inanc Birol, Joseph Connors, Randy Gascoyne, Doug Horsman, Steven J. Jones, and Marco A. Marra. 2009. "Circos: An Information Aesthetic for Comparative Genomics." *Genome Research* 19: 1639–1645.

Kuhn, Thomas S. 1996. *The Structure of Scientific Revolutions*. Chicago, IL: University of Chicago Press.

Lamport, Leslie, and A. LaTEX. 1986. *Document Preparation System*. Reading, MA: Addison-Wesley.

Lander, Eric S., Lauren M. Linton, Bruce Birren, Chad Nusbaum, Michael C. Zody, Jennifer Baldwin, Keri Devon, et al. 2001. "Initial Sequencing and Analysis of the Human Genome." *Nature* 409, no. 6822 (February 15): 860–921.

Lanier, Jaron. 2013. *Who Owns the Future?* London: Allen Lane.

Lantz, Brett. 2013. *Machine Learning with R.* Birmingham: Packt Publishing.

Larsen, Jeff. 2012. "How ProPublica's Message Machine Reverse Engineers Political Microtargeting." Accessed August 28, 2014. http://www.propublica.org/nerds/item/how-propublicas-message-machine-reverse-engineers-political-microtargeting.

Larson, Roland E. 1996. *Elementary Linear Algebra.* 3rd ed. In collaboration with Bruce H. Edwards. Lexington, MA: DC Heath.

Lash, Scott. 2007. "Power After Hegemony: Cultural Studies in Mutation?" *Theory, Culture & Society* 24 (3): 55–78.

Latour, Bruno. 1993. *We Have Never Been Modern.* New York; London: Harvester Wheatsheaf.

Lazzarato, Maurizio. 2014. *Signs and Machines: Capitalism and the Production of Subjectivity.* Cambridge, MA: Semiotext(e).

Le, Quoc V., Marc'Aurelio Ranzato, Rajat Monga, Matthieu Devin, Kai Chen, Greg S. Corrado, Jeff Dean, and Andrew Y. Ng. 2011. "Building High-Level Features Using Large Scale Unsupervised Learning" (December 28). arXiv: 1112.6209.

LeCun, Yann, Bernhard Boser, John S. Denker, Donnie Henderson, Richard E. Howard, Wayne Hubbard, and Lawrence D. Jackel. 1989. "Backpropagation Applied to Handwritten Zip Code Recognition." *Neural Computation* 1 (4): 541–551.

LeCun, Yann, and Corinna Cortes. 2012. "MNIST Handwritten Digit Database, Yann LeCun, Corinna Cortes and Chris Burges." Accessed June 24, 2013. http://yann.lecun.com/exdb/mnist/.

Lee, D. D., and H. S. Seung. 1999. "Learning the Parts of Objects by Non-Negative Matrix Factorization." *Nature* 401, no. 6755 (October 21): 788–791. pmid: 10548103.

Leonelli, Sabina. 2014. "What Difference Does Quantity Make? On the Epistemology of Big Data in Biology." *Big Data & Society* 1 (1): 1–11.

Levine, Sergey, Peter Pastor, Alex Krizhevsky, and Deirdre Quillen. 2016. "Learning Hand-Eye Coordination for Robotic Grasping with Deep Learning and Large-Scale Data Collection" (March 7). arXiv: 1603.02199 [cs]

Levy, Steven. 2016. "How Google Is Remaking Itself as a 'Machine Learning First' Company—Backchannel." June 32. Accessed June 27, 2016. https://backchannel.com/how-google-is-remaking-itself-as-a-machine-learning-first-company-ada63defcb70#. fj3u7o3t2.

Lury, Celia, Luciana Parisi, and Tiziana Terranova. 2012. "Introduction: The Becoming Topological of Culture." *Theory, Culture & Society* 29 (4-5): 3–35.

Lynch, Michael. 1993. *Scientific Practice and Ordinary Action: Ethnomethodology and Social.* Cambridge: Cambridge University Press.

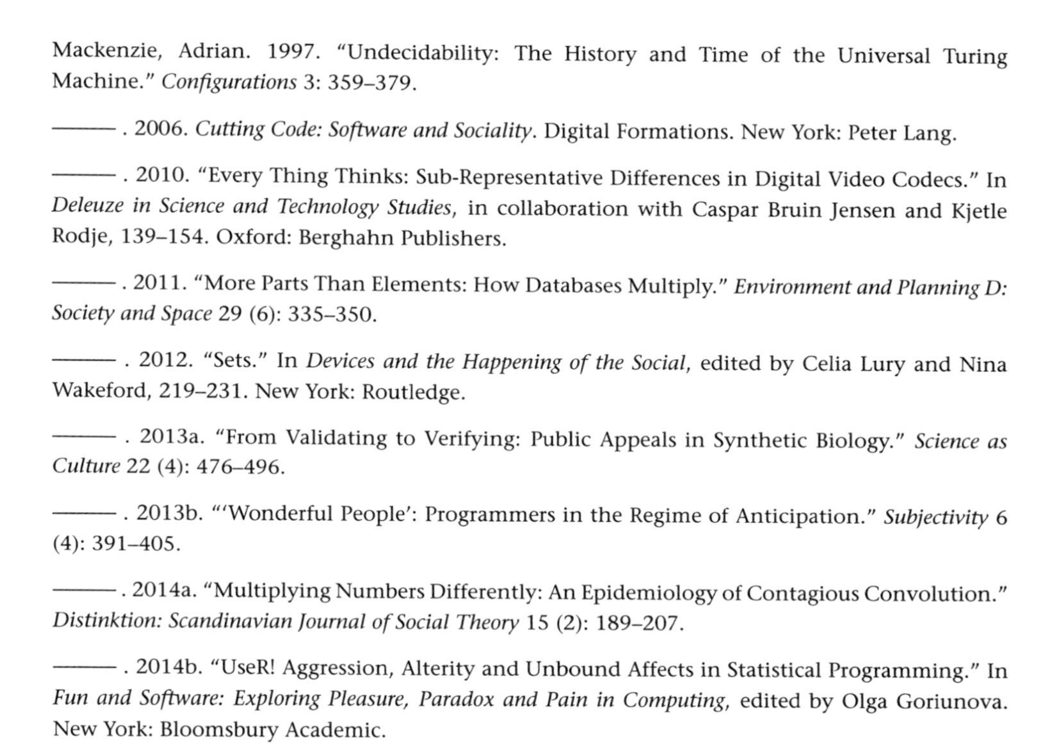

Mackenzie, Adrian. 1997. "Undecidability: The History and Time of the Universal Turing Machine." *Configurations* 3: 359–379.

———. 2006. *Cutting Code: Software and Sociality*. Digital Formations. New York: Peter Lang.

———. 2010. "Every Thing Thinks: Sub-Representative Differences in Digital Video Codecs." In *Deleuze in Science and Technology Studies*, in collaboration with Caspar Bruin Jensen and Kjetle Rodje, 139–154. Oxford: Berghahn Publishers.

———. 2011. "More Parts Than Elements: How Databases Multiply." *Environment and Planning D: Society and Space* 29 (6): 335–350.

———. 2012. "Sets." In *Devices and the Happening of the Social*, edited by Celia Lury and Nina Wakeford, 219–231. New York: Routledge.

———. 2013a. "From Validating to Verifying: Public Appeals in Synthetic Biology." *Science as Culture* 22 (4): 476–496.

———. 2013b. "'Wonderful People': Programmers in the Regime of Anticipation." *Subjectivity* 6 (4): 391–405.

———. 2014a. "Multiplying Numbers Differently: An Epidemiology of Contagious Convolution." *Distinktion: Scandinavian Journal of Social Theory* 15 (2): 189–207.

———. 2014b. "UseR! Aggression, Alterity and Unbound Affects in Statistical Programming." In *Fun and Software: Exploring Pleasure, Paradox and Pain in Computing*, edited by Olga Goriunova. New York: Bloomsbury Academic.

———. 2016. "Distributive Numbers: A Post-Demographic Perspective on Probability." In *Empirical Baroque*, edited by John Law and Evelyn Ruppert. Manchester, UK: Mattering Press.

Mackenzie, Adrian, Matthew Fuller, Andrew Goffey, Richard Mills, and Stuart Sharples. 2016. "Code Repositories as Expressions of Urban Life." In *Code and the City*, edited by Rob Kitchin. London: Routledge.

Mackenzie, Adrian, Richard Mills, Stuart Sharples, Matthew Fuller, and Andrew Goffey. 2015. "Digital Sociology in the Field of Devices." In *Handbook of Sociology of the Arts and Culture*, edited by Mike Savage and Laurie Hanquinet. London; New York: Routledge.

Mackenzie, Adrian, and Simon Monk. 2004. "From Cards to Code: How Extreme Programming Re-Embodies Programming as a Collective Practice." *Computer Supported CooperativeWork (CSCW)* 13 (1): 91–117.

Madrigal, Alexis C. 2014. "How Netflix Reverse Engineered Hollywood." Accessed August 28, 2014. http://www.theatlantic.com/technology/archive/2014/01/how-netflix-reverse-engineered-hollywood/282679/.

Malley, James D., Karen G. Malley, and Sinisa Pajevic. 2011. *Statistical Learning for Biomedical Data*. 1st ed. Cambridge: Cambridge University Press.

Manning, Christopher D., Prabhakar Raghavan, and Hinrich Schüijtze. 2008. *Introduction to Information Retrieval*. 1st ed. Cambridge: Cambridge University Press.

Marchese, Francis T. 2013. "Tables and Early Information Visualization." In *Knowledge Visualization Currents*, edited by Francis T. Marchese and Ebad Banissi, 35–61. London: Springer, January 1.

Markoff, John. 2012. "In a Big Network of Computers, Evidence of Machine Learning." *The New York Times* (June 25).

Maron, M. E. 1961. "Automatic Indexing: An Experimental Inquiry." *Journal of the Association for Computing Machinery* 8: 404–417.

Marx, Karl. 1986. *Capital: A Critique of Political Economy. The Process of Production of Capital.* Moscow: Progress.

Mason, H. 2012. "Hilary Mason—Machine Learning for Hackers." Accessed July 6, 2012. http://vimeo.com/43547079.

Massumi, Brian. 2002. *Parables for the Virtual*. Durham, NC: Duke University Press.

Matloff, Norman S. 2011. *Art of R programming*. San Francisco: No Starch Press.

Mayer-Schönberger, Viktor, and Kenneth Cukier. 2013. *Big Data: A Revolution That Will Transform How We Live, Work, and Think*. Boston: Eamon Dolan/Houghton Mifflin Harcourt.

McClelland, James L., and David E. Rumelhart. 1986. *Parallel Distributed Processing. Explorations in the Microstructure of Cognition*. Vol. 1. Cambridge; London: MIT Press.

McKinney, Wes. 2012. *Python for Data Analysis: Data Wrangling with Pandas, NumPy, and IPython*. Sebastapol, CA: O'Reilly & Associates Inc.

McMillan, Robert. 2013. "How Google Retooled Android With Help From Your Brain." February 18. Accessed August 4, 2015. http://www.wired.com/2013/02/android-neural-network/.

McNally, Ruth, Adrian Mackenzie, Jennifer Tomomitsu, and Allison Hui. 2012. "Understanding the 'Intensive' in 'Data Intensive Research': Data Flows in Next Generation Sequencing and Environmental Networked Sensors." *International Journal of Digital Curation* 7 (1): 81–94.

Meza, Juan C. 2010. "Steepest Descent." *Wiley Interdisciplinary Reviews: Computational Statistics* 2 (6): 719–722.

Minsky, Marvin, and Seymour Papert. 1969. *Perceptron: An Introduction to Computational Geometry*. Cambridge, MA: MIT Press.

Mitchell, Tom M. 1997. *Machine Learning*. New York: McGraw-Hill.

Mnih, Volodymyr, Koray Kavukcuoglu, David Silver, Alex Graves, Ioannis Antonoglou, Daan Wierstra, and Martin Riedmiller. 2013. "Playing Atari With Deep Reinforcement Learning" (December 19). arXiv: 1312.5602 [cs].

Mnih, Volodymyr, Koray Kavukcuoglu, David Silver, Andrei A. Rusu, Joel Veness, Marc G. Bellemare, Alex Graves, et al. 2015. "Human-Level Control Through Deep Reinforcement Learning." Nature 518, no. 7540 (February 26): 529–533.

Mohamed, Abdel-rahman, Tara N. Sainath, George Dahl, Bhuvana Ramabhadran, Geoffrey E. Hinton, and Michael A. Picheny. 2011. "Deep Belief Networks Using Discriminative Features for Phone Recognition," 5060–5063. IEEE, May.

Mohr, John W., and Petko Bogdanov. 2013. "Introduction—Topic Models: What They Are and Why They Matter." *Poetics* 41, no. 6 (December): 545–569.

Mol, Annemarie. 2003. *The Body Multiple: Ontology in Medical Practice*. Durham, NC: Duke University Press.

Montfort, Nick, and Ian Bogost. 2009. *Racing the Beam: The Atari Video Computer System*. Cambridge, MA: MIT Press, January 9.

Moore, David S. 2009. *The Basic Practice of Statistics*. 5th edition. New York; London: W. H. Freeman.

Morgan, James N., and John A. Sonquist. 1963. "Problems in the Analysis of Survey Data, and a Proposal." *Journal of the American Statistical Association* 58 (302): 415–434.

Morton, Timothy. 2013. *Hyperobjects: Philosophy and Ecology After the End of the World*. Minneapolis: University of Minnesota Press.

Muenchen, Robert A. 2014. "The Popularity of Data Analysis Software." Accessed September 2, 2015. http://r4stats.com/articles/popularity/.

Munster, Anna. 2013. *An Aesthesia of Networks: Conjunctive Experience in Art and Technology*. Cambridge, MA: MIT Press.

Myers, Eugene W., Granger G. Sutton, Art L. Delcher, Ian M. Dew, Dan P. Fasulo, Michael J. Flanigan, Saul A. Kravitz, Clark M. Mobarry, Knut H. J. Reinert, Karin A. Remington, et al. 2000. "A Whole-Genome Assembly of Drosophila." *Science* 287 (5461): 2196–2204.

National Security Agency. 2012. "SKYNET: Courier Detection via Machine Learning." Accessed October 29, 2015. https://theintercept.com/document/2015/05/08/skynet-courier/.

NCBI. 2016. "Homo Sapiens Chromosome 15, GRCh38.p7 Primary Assembly" (June 6).

Neyland, Daniel. 2015. "On Organizing Algorithms." *Theory, Culture & Society* 32, no. 1 (January 1): 119–132.

Ng, Andrew. 2008a. *Lecture 1 | Machine Learning (Stanford)*. Accessed January 13, 2013. https://www.youtube.com/watch?v=UzxYlbK2c7E&feature=youtube_gdata_player.

———. 2008b. *Lecture 13 | Machine Learning (Stanford)*. Accessed January 15, 2013. https://www.youtube.com/watch?v=LBtuYU-HfUg&feature=youtube_gdata_player.

———. 2008c. *Lecture 6 | Machine Learning (Stanford)*. Accessed January 15, 2013. https://www.youtube.com/watch?v=qyyJKd-zXRE&feature=youtube_gdata_player.

———. 2008d. *Lecture 7 | Machine Learning (Stanford)*. Accessed January 15, 2013. https://www.youtube.com/watch?v=s8B4A5ubw6c&feature=youtube_gdata_player.

———. 2008e. *Lecture 10 | Machine Learning (Stanford)*. Accessed January 15, 2013. https://www.youtube.com/watch?v=0kWZoyNRxTY&feature=youtube_gdata_player.

———. 2008f. *Lecture 2 | Machine Learning (Stanford)*. Accessed January 15, 2013. https://www.youtube.com/watch?v=5u4G23_OohI&feature=youtube_gdata_player.

———. 2008g. *Lecture 3 | Machine Learning (Stanford)*. Accessed January 15, 2013. https://www.youtube.com/watch?v=HZ4cvaztQEs&feature=youtube_gdata_player.

———. 2008h. *Lecture 9 | Machine Learning (Stanford)*. Accessed January 15, 2013. https://www.youtube.com/watch?v=tojaGtMPo5U&feature=youtube_gdata_player.

NIST. 2012. "Gallery of Distributions." Accessed September 21, 2012. http://www.itl.nist.gov/div898/handbook/eda/section3/eda366.htm.

Olazaran, Mikel. 1996. "A Sociological Study of the Official History of the Perceptrons Controversy." *Social Studies of Science* 26, no. 3 (January 8): 611–659.

Pariser, Eli. 2011. *The Filter Bubble: What the Internet is Hiding From You*. London: Penguin.

Parisi, Luciana. 2013. *Contagious Architecture: Computation, Aesthetics and Space*. Cambridge, MA: MIT Press.

Parry, R. M., W. Jones, T. H. Stokes, J. H. Phan, R. A. Moffitt, H. Fang, L. Shi, A. Oberthuer, M. Fischer, W. Tong, et al. 2010. "K-Nearest NeighborModels for Microarray Gene Expression Analysis and Clinical Outcome Prediction." *The Pharmacogenomics Journal* 10 (4): 292–309.

Pasquinelli, Matteo. 2014. "Italian Operaismo and the Information Machine." *Theory, Culture & Society* (February 2): 1–20.

———. 2015. *Anomaly Detection: The Mathematization of the Abnormal in the Metadata Society*. Berlin: Transmediale.

Pearson, Karl. 1901. "LIII. On Lines and Planes of Closest Fit to Systems of Points in Space." *Philosophical Magazine Series 6* 2, no. 11 (November 1): 559–572.

Pedregosa, F., G. Varoquaux, A. Gramfort, V. Michel, B. Thirion, O. Grisel, M. Blondel, et al. 2011. "Scikit-learn: Machine Learning in Python." *Journal of Machine Learning Research* 12: 2825–2830.

Peirce, Charles Sanders. 1992. *The Essential Peirce: 1867-1893 v. 1: Selected Philosophical Writings*. New York: John Wiley & Sons.

———. 1998. *The Essential Peirce—Volume 2: Selected Philosophical Writings: (1893-1913) v. 2*. Indianapolis: Indiana University Press.

Perceptron. 2013. In *Wikipedia, the Free Encyclopedia*, by Wikipedia.

Perez, Fernando, and Brian E. Granger. 2007. "IPython: A System for Interactive Scientific Computing." *Computing in Science & Engineering* 9 (3): 21–29.

Petrova, Svetlana S., and Alexander D. Solov'ev. 1997. "The Origin of the Method of Steepest Descent." *Historia Mathematica* 24, no. 4 (November): 361–375.

Pevzner, Pavel A., Haixu Tang, and Michael S. Waterman. 2001. "An Eulerian Path Approach to DNA Fragment Assembly." *Proceedings of the National Academy of Sciences* 98, no. 17 (August 14): 9748–9753. pmid: 11504945.

Quinlan, J. Ross. 1986. "Induction of Decision Trees." *Machine Learning* 1 (1): 81–106.

Quinlan, John Ross. 1993. *C4. 5: Programs for Machine Learning*. Vol. 1. San Francisco, CA: Morgan Kaufmann.

Rabiner, Lawrence. 1989. "A Tutorial on Hidden Markov Models and Selected Applications in Speech Recognition." *Proceedings of the IEEE* 77 (2): 257–286.

Ramaswamy, Sridhar, Pablo Tamayo, Ryan Rifkin, Sayan Mukherjee, Chen-Hsiang Yeang,Michael Angelo, Christine Ladd, Michael Reich, Eva Latulippe, Jill P. Mesirov, et al. 2001. "Multiclass Cancer Diagnosis Using Tumor Gene Expression Signatures." *Proceedings of the National Academy of Sciences* 98 (26): 15149–15154.

RexerAnalytics. 2015. "Rexer Analytics 7th Annual Data Miner Survey—2015." Accessed May 9, 2011. http://www.rexeranalytics.com/Data-Miner-Survey-2015-Intro.html.

Richert, Willi, and Luis Pedro Coelho. 2013. *Building Machine Learning Systems With Python*. Birmingham, AL: Packt Publishing.

Ripley, Brian. 1996. *Pattern Recognition and Neural Networks. 1996*. Cambridge; New York: Cambridge University Press.

———. 2014. *Tree: Classification and Regression Trees*. R package version 1.0-37. https://CRAN.R-project.org/package=tree.

Robinson, Derek. 2008. "Function." In *Software Studies: A Lexicon*, edited by M. Fuller, 101–110. Cambridge, MA: The MIT Press.

Rose, N. 2009. "Normality and Pathology in a Biomedical Age." *Sociological Review* 57: 66–83.

Rosenblatt, F. 1958. "The Perceptron: A Probabilistic Model for Information Storage and Organization in the Brain." *Psychological Review* 65 (6): 386–408.

Rumelhart, David E., Geoffrey E. Hinton, and Ronald J. Williams. 1985. *Learning Internal Representations by Error Propagation*. DTIC Document.

———. 1986. "Learning Representations by Back-Propagating Errors." *Nature* 323, no. 6088 (October 9): 533–536.

Russell, Matthew A. 2011. *Mining the Social Web*. Sebastopol, CA: O'Reilly.

Savage, M. 2009. "Contemporary Sociology and the Challenge of Descriptive Assemblage." *European Journal of Social Theory* 12 (1): 155.

Schutt, Rachel, and Cathy O'Neil. 2013. *Doing Data Science*. Sebastopol, CA: O'Reilly & Associates Inc.

Segaran, Toby. 2007. *Programming Collective Intelligence: Building Smart Web 2.0 Applications*. Sebastapol CA: O'Reilly.

Silver, David, Aja Huang, Chris J. Maddison, Arthur Guez, Laurent Sifre, George Van Den Driessche, Julian Schrittwieser, Ioannis Antonoglou, Veda Panneershelvam, Marc Lanctot, et al. 2016. "Mastering the Game of Go With Deep Neural Networks and Tree Search." *Nature* 529 (7587): 484–489.

Slikker, W., Jr. 2010. "Of Genomics and Bioinformatics." *The Pharmacogenomics Journal* 10, no. 4(August): 245–246. pmid: 20676063.

Smith, Marquard. 2013. "Theses on the Philosophy of History: The Work of Research in the Age of Digital Searchability and Distributability." *Journal of Visual Culture* 12, no. 3(December 1): 375–403.

Stamey, Thomas A., Mitchell Caldwell, John McNeal, Rosalie Nolley, Marci Hemenez, and Joshua Downs. 2004. "The Prostate Specific Antigen Era in the United States Is Over for Prostate Cancer: What Happened in the Last 20 Years?" *The Journal of Urology* 172 (4): 1297–1301.

Stamey, Thomas A., John N. Kabalin, John E. McNeal, Iain M. Johnstone, F. Freiha, Elise A. Redwine, and Norman Yang. 1989. "Prostate Specific Antigen in the Diagnosis and Treatment of Adenocarcinoma of the Prostate. II. Radical Prostatectomy Treated Patients." *The Journal of Urology* 141 (5): 1076–1083.

Steinberg, Dan, and Phillip Colla. 2009. "CART: Classification and Regression Trees." *The Top Ten Algorithms in Data Mining*, 179–201.

Stengers, Isabelle. 2000. *The Invention of Modern Science. Theory Out of Bounds*. Volume 19. Minneapolis; London: University of Minnesota Press.

———. 2005. "Deleuze and Guattari's Last Enigmatic Message." *Angelaki* 10 (1): 151–167.

———. 2008. "Experimenting With Refrains: Subjectivity and the Challenge of Escaping Modern Dualism." *Subjectivity* 22, no. 1 (May): 38–59.

———. 2011. *Cosmopolitics II*. Translated by Robert Bononno. Minneapolis: University of Minnesota Press, September 26.

Stevens, Hallam. 2011. "Coding Sequences: A History of Sequence Comparison Algorithms as a Scientific Instrument." *Perspectives on Science* 19 (3): 263–299.

———. 2013. *Life Out of Sequence: A Data-Driven History of Bioinformatics*. Chicago; London: University of Chicago Press.

Stigler, Stephen M. 1986. *The History of Statistics: The Measurement of Uncertainty Before 1900*. Cambridge, MA: Harvard University Press.

——— . 2002. *Statistics on the Table: The History of Statistical Concepts and Methods*. Cambridge, MA: Harvard University Press.

Stone, Mervyn. 1974. "Cross-Validatory Choice and Assessment of Statistical Predictions." *Journal of the Royal Statistical Society. Series B (Methodological)*, 111–147. JSTOR 2984809.

Suchman, Lucy. 2006. *Human and Machine Reconfigurations: Plans and Situated Actions*. 2nd ed. Cambridge: Cambridge University Press, December 4.

Suchman, Lucy A. 1987. *Plans and Situated Actions: The Problem of Human-Machine Communication*. Cambridge: Cambridge University Press.

Suchman, Lucy A. and Randall H. Trigg. 1992. "Artificial Intelligence as Craftwork." In *Understanding Practice: Perspectives on Activity and Context*, edited by Seth Chaiklin and Jean Lave, 144–178. Cambridge; New York: Cambridge University Press.

Sunder Rajan, Kaushik. 2006. *Biocapital: The Constitution of Postgenomic Life*. Durham, NC: Duke University Press.

Tarde, Gabriel de. 1902. *Psychologie Économique*. Paris: F. Alcan.

Teetor, Paul. 2011. *R Cookbook*. Sebastopol, CA: O'Reilly.

Thacker, Eugene. 2005. *The Global Genome: Biotechnology, Politics, and Culture*. Cambridge, MA: MIT Press.

Therneau, Terry, Beth Atkinson, and Brian Ripley. 2015. *Rpart: Recursive Partitioning and Regression Trees*. http://CRAN.R-project.org/package=rpart.

Thrun, Sebastian, Mike Montemerlo, Hendrik Dahlkamp, David Stavens, Andrei Aron, James Diebel, Philip Fong, John Gale, Morgan Halpenny, and Gabriel Hoffmann. 2006. "Stanley: The Robot That Won the DARPA Grand Challenge." *Journal of Field Robotics* 23 (9): 661–692.

Tibshirani, Robert. 1996. "Regression Shrinkage and Selection via the Lasso." *Journal of the Royal Statistical Society. Series B* (*Methodological*): 267–288. JSTOR 2346178.

Totaro, Paolo, and Domenico Ninno. 2014. "The Concept of Algorithm as an Interpretative Key of Modern Rationality." *Theory, Culture & Society*, 29–49.

Tuv, E., A. Borisov, G. Runger, and K. Torkkola. 2009. "Feature Selection With Ensembles, Artificial Variables, and Redundancy Elimination." *The Journal of Machine Learning Research* 10:1341–1366.

Valiant, Leslie G. 1984. "A Theory of the Learnable." *Communications of the ACM* 27 (11): 1134–1142.

Van Dijck, José. 2012. "Facebook and the Engineering of Connectivity: A Multi-Layered Approach to Social Media Platforms." *Convergence: The International Journal of Research into New Media Technologies* 19 (2): 141–155.

Vance, Ashlee. 2011. "This Tech Bubble Is Different." *Business Week: Magazine* (April 14).

Vapnik, Vladimir. 1999. *The Nature of Statistical Learning Theory.* 2nd. ed. New York: Springer.

Vapnik, Vladimir N., and A. Ya Chervonenkis. 1971. "On the Uniform Convergence of Relative Frequencies of Events to Their Probabilities." *Theory of Probability & Its Applications* 16 (2): 264–280.

Venables, William N., and Brian D. Ripley. 2002. *Modern Applied Statistics With S.* New York: Springer.

Venter, J. Craig, Mark D. Adams, Eugene W. Myers, Peter W. Li, Richard J. Mural, Granger G. Sutton, Hamilton O. Smith, et al. 2001. "The Sequence of the Human Genome." Science 291, no. 5507 (February 16) 1304–1351. pmid: 11181995.

Virno, Paolo. 2004. *A Grammar of the Multitude: For an Analysis of Contemporary Forms of Life.* Semiotext(e) foreign agents series. Los Angeles: Semiotext(e).

Warner, Homer R., Alan F. Toronto, L. George Veasey, and Robert Stephenson. 1961. "A Mathematical Approach to Medical Diagnosis: Application to Congenital Heart Disease." *Jama* 177 (3): 177–183.

Warner, Homer R., Alan F. Toronto, and L. George Veasy. 1964. "Experience With Baye's Theorem for Computer Diagnosis of Congenital Heart Disease*." *Annals of the New York Academy of Sciences* 115, no. 2 (July 1) 558–567.

Wasserman, Larry. 2003. *All of Statistics: A Concise Course in Statistical Inference.* New York: Springer.

Whitehead, Alfred North. 1956. *Modes of Thought; Six Lectures Delivered in Wellesley College, Massachusetts, and Two Lectures in the University of Chicago.* New York: Cambridge University Press.

———. 1960. *Process and Reality, an Essay in Cosmology.* New York: Macmillan.

Wiener, Norbert. 1961. *Cybernetics, or, Control and Communication in the Animal and the Machine.* 2nd ed. Cambridge, MA: MIT Press.

Wikibooks. 2013. "R Programming—Wikibooks, Open Books for an Open World." Accessed June 27, 2013. http://en.wikibooks.org/wiki/R_Programming.

Wilf, Eitan. 2013. "Toward an Anthropology of Computer-Mediated, Algorithmic Forms of Sociality." *Current Anthropology* 54, no. 6 (December 1) 716–739. JSTOR 10.1086/673321.

Wilson, Elizabeth A. 2010. *Affect and Artificial Intelligence.* Seattle: University of Washington Press.

Witten, Ian H., and Eibe Frank. 2005. *Data Mining: Practical Machine Learning Tools and Techniques.* New York: Morgan Kaufmann.

Wu, X., V. Kumar, J. Ross Quinlan, J. Ghosh, Q. Yang, H. Motoda, G. J McLachlan, A. Ng, B. Liu, P. S Yu, et al. 2008. "Top 10 Algorithms in Data Mining." *Knowledge and Information Systems* 14 (1): 1–37.

Xie, Yihui. 2013. "Knitr: A General-Purpose Package for Dynamic Report Generation in R." *R Package Version* 1.

Xie, Yihui, and J. J. Allaire. 2012. *New Tools for Reproducible Research with R*. Presented at the useR Conference, Warwick, UK.

Zare, Douglas. 2012. "Difference Between Logistic Regression and Neural Networks—Cross Validated." Accessed May 26, 2015. http://stats.stackexchange.com/questions/43538/difference-between-logistic-regression-and-neural-networks.

Zerbino, D. R., B. Paten, and D. Haussler. 2012. "Integrating Genomes." *Science* 336, no. 6078 (April 13): 179–182.

Index